MW00398655

ON-SITE FOODSERVICE MANAGEMENT

ON-SITE FOODSERVICE MANAGEMENT
A Best Practices Approach

DENNIS REYNOLDS

JOHN WILEY & SONS, INC.

This book is printed on acid-free paper.♾

Copyright © 2003 by Dennis Reynolds. All rights reserved.

Published by John Wiley & Sons, Inc., Hoboken, New Jersey.
Published simultaneously in Canada.

No part of this publication may be reproduced, stored in a retrieval system or transmitted in any form or by any means, electronic, mechanical, photocopying, recording, scanning or otherwise, except as permitted under Section 107 or 108 of the 1976 United States Copyright Act, without either the prior written permission of the publisher, or authorization through payment of the appropriate per-copy fee to the Copyright Clearance Center, Inc., 222 Rosewood Drive, Danvers, MA 01923, (978) 750-8400, fax (978) 750-4470, or on the web at www.copyright.com. Requests to the publisher for permission should be addressed to the Permissions Department, John Wiley & Sons, Inc., 111 River Street, Hoboken, NJ 07030, (201) 748-6011, fax (201) 748-6008, e-mail: permcoordinator@wiley.com.

Limit of Liability/Disclaimer of Warranty: While the publisher and author have used their best efforts in preparing this book, they make no representations or warranties with respect to the accuracy or completeness of the contents of this book and specifically disclaim any implied warranties of merchantability or fitness for a particular purpose. No warranty may be created or extended by sales representatives or written sales materials. The advice and strategies contained herein may not be suitable for your situation. You should consult with a professional where appropriate. Neither the publisher nor author shall be liable for any loss of profit or any other commercial damages, including but not limited to special, incidental, consequential, or other damages.

For general information on our other products and services or for technical support, please contact our Customer Care Department within the United States at (800) 762-2974, outside the United States at (317) 572-3993 or fax (317) 572-4002.

Wiley also publishes its books in a variety of electronic formats. Some content that appears in print may not be available in electronic books. For more information about Wiley products, visit our web site at www.wiley.com.

Library of Congress Cataloging-in-Publication Data:

Reynolds, Dennis E.
  Onsite foodservice management : a best practices approach / Dennis
Reynolds.
    p. ; cm.
Includes bibliographical references.
  ISBN 0-471-34543-1 (alk. paper)
  1. Hospitals—Food service.
  [DNLM: 1. Food Service, Hospital—organization & administration. WX
168 R462o 2003] I. Title.
  RA975.5.D5 R49 2003
  362.1'76—dc21   2002005426

Printed in the United States of America.

10 9 8 7 6 5 4 3 2 1

As always, I am deeply appreciative of my wife, Julia, and children, Sydney and Quincy, for their forbearance and love. It is to them that I dedicate this book.

CONTENTS

PREFACE

On-site foodservice, historically referred to as "noncommercial foodservice," has changed considerably over the years. Once considered the poor stepchild of the traditional restaurant industry, it has achieved unprecedented stature, with annual worldwide sales expected to top $230 billion in this decade. Moreover, opportunities for individuals with honed management and leadership skills—both those employed by leading managed-services companies and freelance workers seeking self-operated settings—are at an all-time high owing to the on-site segment's growth and the limited labor pool of qualified individuals.

Along with the segment's evolution, the management of on-site foodservice operations has become highly specialized. Today, managers must possess a keen knowledge of foodservice operations, a prowess at leading and motivating a diverse workforce, and a knack for marketing to a generally limited yet extremely savvy customer base. These managers also need a variety of sources from which to draw new ideas in their unending effort to maximize revenue while minimizing costs.

To this end, the purpose of this book is to provide managers with pioneering examples—best practices—of what is happening in the marketplace. Drawn from the healthcare-foodservice arena, these practices exemplify means of producing the best possible performance. In addition, the discussions surrounding these best practices offer managerial perspectives on relevant issues with the goal of presenting a holistic view of on-site foodservice management.

WHY EMPLOY A BEST PRACTICES APPROACH?

"Best practices," by definition, refers to any practice, know-how, or experience that has proved valuable or effective in a specific setting and that may have applicability in other situations. Thus, a best practice may be specific to a certain organization, a specific set of conditions, and a certain aspect of an operation, but—of greatest importance—it is portable. Ultimately, a best practice designation indicates that, at least in the context of this book, the approach serves as a benchmark and distinguishes the organization that uses it as an exemplar within a highly competitive peer group.

Best practices typically benefit organizations indirectly through lower operating costs, including labor expenses, increased revenue, and enhanced utilization of human capital while directly increasing short- or long-term profitability. For an individual manager, however, the study of best practices from the same industry allows him or her to integrate proven practices with minimal trial and error—it lessens the risk. Of still greater importance, the study of best practices can spark new ideas about the possibilities for a manager's specific operation.

In order to identify best practices with the greatest utility, we conducted research for four years that included survey methods, interviews, and site visits. In most instances, the goal was to identify a best practice given specific situational contexts. The research included more than 3,000 on-site operations on several continents. The result is an impressive list of 22 best practices, each of which exemplifies a high degree of innovation, effectiveness, and application to a broad range of on-site foodservice operations.

WHY FOCUS ON THE HEALTHCARE SEGMENT?

As managers who have worked in healthcare foodservice know, this segment tops the charts with constant pressures to operate efficiently and at the lowest possible cost. To exacerbate such extreme conditions, most healthcare-related foodservice departments operate within organizations whose cultures mandate a systemwide quest for quality in everything from emergency room procedures to employee dining. In addition, this segment of on-site foodservice faces additional challenges in governmental regulations, a customer base (including patients, employees, and visitors to the healthcare center) with diverse dietary and service needs, and a long-standing reputation for low-quality food. After all, who hasn't heard someone refer to hospital food with the implicit message that it is bland, tasteless, and unsatisfying? What better segment is there, then, to profile the best practices in on-site foodservice? Finally, foodservice in the healthcare arena is unquestionably challenging, but it is also resilient. It is continually generating many solutions to the problems common to on-site foodservice operators around the globe.

TEXT FORMATION

The book is divided into four parts. Part I is an overview, which includes a history of on-site foodservice, a discussion of organizational structure and the foodservice operation's role, and—last but not least—a summary of different operational configurations including different production approaches. Part II focuses on external customers, profiling the unique set of constituencies that on-site foodservice must serve. These include employees, patients (in the healthcare segment), and visitors, as well as the more traditional markets serviced through catering efforts. Part III considers operations with a focus on the internal customers and the systems used to produce food and deliver services. Included in this part are unique features such as methods for defining, measuring, and enhancing productivity (Chapter 9 is dedicated to this single area). Finally, Part IV delves into trends and challenges. Specifically, it investigates the whole notion of quality. It looks at the new frontier of senior dining and the possibilities and challenges of melding foodservice with other support services in an on-site setting. The part concludes with an exploration of new technologies that will likely shape the future of on-site foodservice in the decades to come.

WHO SHOULD READ THIS BOOK?

This book is geared to managers in any segment of on-site foodservice interested in improving their foodservice operations. The discussion and accompanying examples are intended to provide new ideas, build on existing systems, and enhance overall operations for managers young or and old, inexperienced and well versed. The book also offers considerable utility for students wishing to understand the challenges of on-site foodservice. Appropriate for upperclass and graduate students in hospitality programs, the book builds on information found in traditional restaurant-management textbooks. This book requires that students extend their view of restaurant operations, however, to understand the unique challenges of this dynamic segment. In order to provoke thought and stimulate discussion, whether in the manager's office or the classroom, each chapter concludes with a list of questions surrounding the topics presented.

Today's customers—whether in healthcare foodservice or any other on-site foodservice segment—are becoming more and more educated and demanding. Paralleling this trend, developments in market segmentation and penetration, service delivery (and the too often associated need for service recovery), and distribution increasingly require managers who possess unprecedented acumen in their respective sectors of on-site foodservice, with foresight, new tools, and knowledge of possible alternatives. This book is intended to facilitate management development and analytical thought, which should in turn help today's best managers expand their toolkits and hone their current operations.

RECOGNITION

In closing, I must acknowledge and thank the many expert managers who have shared their best practices with me. In addition, I thank the many managers, some of whom I have known for many years, who offered their input. Many of these individuals are still in on-site operations in different segments, settings, and countries, living the reality of the challenges addressed here. Others have left the foodservice director's office for other opportunities. In either case, this book reflects a wide range of experiences and a general orientation toward exceeding the customer's hospitality expectations. The managers I have met over the years, and particularly those whose operations are featured here, certainly have exceeded mine as I have watched so many shape unwieldy food outlets into impressive on-site foodservice operations that rival the latest in profitable chain restaurants. The best practices featured in this book and the managers who have developed them are evidence of this impressive expertise.

This book also reflects the behind-the-scenes efforts of many others who have helped me surmount logistical challenges. JoAnna Turtletaub of John Wiley & Sons repeatedly demonstrated the patience and support of a saint. Lisa Macleod was instrumental in helping with the less glamorous aspects of the text. Bill Barnett was amazing in keeping my logical arguments in check. My students were invaluable in helping me with delivery and packaging. And finally, my family and friends were always there for me, whether with a glass of wine to celebrate a chapter's end or a receptive ear to absorb the frustration of an unfinished section. I humbly extend my gratitude to all of these individuals; this book is the fruit of your support, skills, and caring.

OVERVIEW

The three chapters in Part I provide a broad overview of on-site foodservice. Beginning with the early evolution of foodservice, Chapter 1 serves as a backdrop for a discussion of what customers expect and what challenges accompany these expectations. The chapter describes the different segments of on-site foodservice from a historical perspective; it also defines the industry in greater detail in terms of both dominant market leaders and differences between independent and self-operated units.

Chapter 2 builds on the discussion of on-site foodservice's evolution by presenting the changing picture of the typical organizational structure of an entity offering on-site foodservice as well as the foodservice department's role within such an organization. It describes the shifting emphasis toward the end user and the evolving needs of customers. The chapter ends by looking more closely at an organization's new perspective on support services—namely, foodservice—and how companies view foodservice operations. This is particularly important given the already challenging task of producing consistently high-quality food products at an attractive price.

Chapter 3 explores different operational configurations that are found today in on-site foodservice, with particular attention to the differences between centralized and decentralized production systems. The discussion then turns to the topic of delivery systems and how these can best be utilized, depending on an organization's needs and configuration. Finally, the chapter, and Part I, conclude by considering different service styles and how these can be embraced to truly exceed customer expectations.

CHAPTER 1

ON-SITE FOODSERVICE— AN INTRODUCTION

It is difficult to imagine a time when there were very limited outlets for food and beverages, when the choice was not where to eat but whether there was anywhere to eat besides one's own kitchen. For that matter, it is strange to think of going to a hospital that doesn't have a cafeteria for visitors and employees or a corporate headquarters that doesn't have an eatery for the building's occupants. Even in elementary schools, one expects to see vending machines strategically placed in the corridors and a lunchroom that provides meals for students. This commonplace availability of food and service is, however, a relatively new trend. In fact, it wasn't until the second half of the twentieth century that food was a common offering of businesses, schools, and hospitals.

This chapter begins by formally describing on-site foodservice and reviewing its evolution. It continues with a discussion of the challenges of on-site foodservice, which, not coincidentally, have increased with rising customer expectations. Next, the chapter discusses the industry's unique features and the ramifications of a hospitality business that is at times strongly divided between operations managed by professional firms and self-operated foodservice departments, concluding with a discussion regarding leading managed-services companies.

ON-SITE FOODSERVICE DEFINED

Ever since leaving the cave, we humans have treated food as more than just sustenance. As early as 3500 B.C., humankind began to realize that food is a commodity that can be prepared and sold or exchanged. It was around this time that the first commercial restaurant was opened in ancient Babylonia. It is believed that this was, in essence, an extension of the home where travelers could rest in a spare room and enjoy meals with the resident family.

In addition, chefs were called on very early in history to focus on quantity food production. Even predating commercial applications of foodservice, quantity food production was needed to support such major human feats as the building of the pyramids. Lest they lose their slaves to starvation, the pharaohs mandated that food be provided. Granted, the quality of food was at a very different level from what is considered common today but, nonetheless, this served as the precursor to what is now found around the globe.

From these humble beginnings, a huge business was created. Today, the **on-site foodservice** segment represents approximately $230 billion of an $800 billion global foodservice industry. On-site foodservice is defined as *food outlets in business and industry (B&I), schools, universities and colleges, hospitals, skilled-nursing centers, eldercare centers, correctional facilities, recreation facilities such as stadiums, and childcare centers.*[1] Some have included military and transportation-related foodservice (e.g., airlines) as part of the on-site world. However, these subsegments have become so highly specialized and distant from the primary on-site business that they are now considered separate and distinct. Moreover, specialized services such as the preparation of airline meals and snacks now have more in common with commodity production than with foodservice.

In light of the rampant downsizing of the 1990s, quantitatively segmenting on-site is, at best, a task of approximation, at least for business in the United States. Nonetheless, experts have suggested that for the beginning of the twenty-first century, the breakdown features B&I and the education markets constituting the largest chunks, closely followed by healthcare, as shown in Figure 1.1.

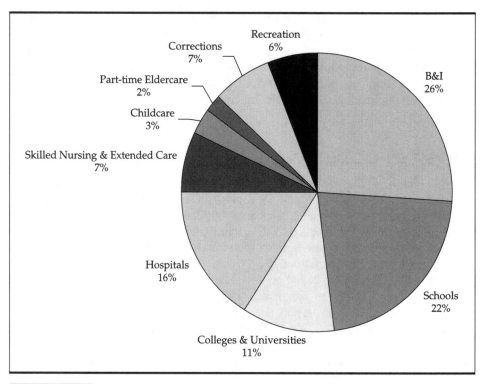

FIGURE 1.1 Breakdown of On-Site Foodservice Sales by Segment.

THE EVOLUTION OF ON-SITE FOODSERVICE

In looking at the evolution of on-site foodservice, each of the segments stands out by virtue of its unique characteristics. Most notably, however, both healthcare and B&I foodservice have deep roots.

FIGURE 1.2 A Copy of the Code of Hammurabi Engraved on a Block of Black Basalt, Circa 1775 B.C. Photo Made Available by the University of Chicago Press.

Healthcare

The history of **foodservice in healthcare** parallels the evolution of hospitals. The first evidence of healthcare and its requisite support services, such as the provision of meals, dates back to 1775 B.C. and can be found in the **Code of Hammurabi,** a collection of the laws and edicts of that famous Babylonian king. It is the earliest legal code known in its entirety (see Figure 1.2). In the code are instructions about the appropriate payment that is due caregivers for care and food given to sick individuals.[2]

In the early Middle Ages, most hospitals were built in conjunction with monasteries and were found at places of pilgrimage or along well-traveled roads. These were operated more for housing and feeding poor and sick travelers than for healing locals. In fact, these, as well as such related structures as leper houses, provided no medical or nursing care other than a solid meal.

By the 1500s, the rise of cities and metropolitan centers sparked the need for more institutionalized forms of healthcare. With this came greater accessibility to foodservice for the sick as well as growth in the relationship between food and healing. Correspondingly, the need for qualified labor in the kitchen became an issue. In England, for example, records indicate that the average wage for a hospital cook in 1552 was £8 per year (about US$11), which was approximately equal to the wage paid for staff providing direct health-related services.[3]

Things changed very little and scientific development in medical approaches to heal patients was minimal until the eighteenth century. Healthcare foodservice, too, remained somewhat rudimentary until this time. Florence Nightingale, at the Institute for Sick Gentlewomen in London, began developing menus that deviated from the bread, beef, and beer that had typified most main meals in hospitals for decades. She also negotiated contracts with food producers at wholesale prices instead of buying food from retailers, changing common practice.[4]

Alexis Soyer, a famous chef of the era, joined Ms. Nightingale in the Crimea and revolutionized hospital kitchens in other ways. He insisted on a permanent staff he could develop as professional cooks; he also served patients himself, thereby facilitating an exchange of information about the food and the care. Notably, he is credited with standardizing hospital menus and developing cooking methods targeted at maximizing the economies of quantity food production while also maximizing the nutritive properties of menu items.

These and many other changes (see Figures 1.3 and 1.4) have contributed to how we experience healthcare today. More recent changes, such as the widespread coverage of hospital costs by insurance, along with shorter stays and increased regulatory guidelines, have resulted in care that is both holistic and technologically complex. Another factor contributing to this complexity is capitated cost structures. **Capitation** refers to reimbursing the healthcare organization on a fixed-per-member-per-month basis to cover all institutional costs for a defined population of members. This is vastly different from the traditional fee-for-service form of reimbursement. Capitation rates are most often determined by the age and sex of the member, since there is a correlation between these factors and utilization of services.

At the same time, foodservice is now widely recognized as a key to patient care in terms of both satisfaction and healing. This is true regardless of the type of health-

FIGURE 1.3 Cayuga Medical Center (Ithaca, New York) Employee Dining Room in Action, Circa 1952. Copyright by Cayuga Medical Center at Ithaca Archives.

care institution, whether providing acute care or offering various degrees of assisted living or extended care.

Business and Industry (B&I)

B&I foodservice dates back almost as far as healthcare. As noted earlier, the pharaohs realized that providing food in the workplace (albeit to slaves) was critical to maintaining labor productivity. While the direct cost of labor was immaterial, the cost of food, based on a cost-benefit analysis looking at potential productivity, was warranted. This concept underlies B&I foodservice throughout history.

In wealthy medieval households, the serving staff often included hundreds of individuals. Since many of these employees were expected to be on call 24 hours a day, it was common to provide them meals. Employee meals were prepared at nontraditional times (just as a modern restaurant might feed its employees in the late

FIGURE 1.4 Cayuga Medical Center's Public Dining Area in 2002. Copyright by Cayuga Medical Center at Ithaca Archives.

afternoon, before the evening rush). The cost of these meals was typically included in the cost of maintaining the household and was integrated into the overall food budget. Since pay for such positions was low (or existed only in the form of a place to live), this benefit of meals was sizable. It is interesting, too, that different employees commonly enjoyed different types of food. Those in more important positions, for example those who directly served the lord of the estate, such as the head chef, enjoyed better meals, while those responsible for more commonplace duties, such as cleaning laundry, were apportioned more modest fare.

Similar arrangements were found during the Middle Ages when apprentices received meals as part of their employ. Like the slaves of past eras, employees of the time worked long hours and required meals to survive. Moreover, since labor was an inexpensive commodity, the level of foodservice provided to the staff was rudimentary (and cheap). Employers were in a position to summarily dismiss employees who complained, without concern for filling the position. It was not unusual for employees involved in large building projects to receive only a scrap of bread and perhaps a scoop of porridge during the course of a workday.

The early years of the Industrial Revolution brought changes not only to the economies of business but also to the utilization of labor. While still arguably viewing employees as a disposable commodity, employers now wanted more than ever to maximize the productivity of their workers. Robert Owen, a British utopian socialist, was one of the earliest to realize the importance of food in maximizing employee productivity. He was an industrial pioneer who believed that by enhancing his workers' environment, he could increase productivity and profit. To this end, he built a separate kitchen in his textile mill dedicated to preparing wholesome employee meals; he also provided a large eating room for employees and their families in his plant in New Lanark, Scotland, in 1805. Cleverly, he charged a nominal price for the meals to make them more valued to the employees.[5] Owen's approach to business was the first example of **subsidized** employee **meals**—meals paid for in part or entirely by the employer as a benefit for employees.

As the supply-demand balance began to shift dramatically during the early years of World War II, Owen's notion of wooing employees through the workplace environment became particularly important in American factories. With competition at a peak, plant managers eagerly complied with blue-collar employees' requests for quality meals served on-site. The notion of largely subsidizing employee meals became commonplace at this time, with employees regarding meals at work as both nutritious and economical.

In the 1960s the white-collar side of B&I grew, and in the early 1970s office-based businesses, seeing the opportunity for increased productivity and job satisfaction, began to provide employee cafeterias or eateries as part of the main office complex. Here, the trade-off between a meal subsidy and the potential productivity gains resulted in the same high-value meals found in factories. Executive suites in company headquarters took the concept even further, adding executive chefs dedicated to preparing haute cuisine for the executive dining room.

Today, subsidized foodservice is more the exception than the norm. In most instances, whether in factories or office complexes, an on-site eatery must be economically self-sufficient and sometimes must operate at a profit. This is extraordinarily challenging, since employees' eating options outside the workplace are today more plentiful than ever. Furthermore, employees still think of employee dining rooms as inexpensive. Thus, operators must compete with new cafes while also offering exceptional value.

Education

As is the case with healthcare foodservice and the evolution of hospitals, **school foodservice** has evolved along with educational systems. This has occurred in two somewhat similar ways—one in schools, and the other in colleges and universities. Until the mid-nineteenth century, school foodservice was unknown save for monastic settings or orphanages that provided meals as part of the greater institution. The early schoolhouse was small and was intended to provide basic education to a wide range of children. Time for education in a child's day was limited because their parents needed them at home. It was simply more productive for children to eat at home.

In the mid-1800s, however, societal norms began to change, as did the need for child labor, owing to the same transformation that affected B&I foodservice. At that time, parents began to value education for their children. With this change, parents and some public officials in Europe during the late 1840s—and a few years later in the United States—launched various foodservice programs as incentives to children to attend school and to parents to send their children to school. Such a perquisite was highly valued by many. In the poor areas of some cities, parents valued the meals over the education. The approach worked.

With the onset of World War I, the need for better nutrition came to the forefront, as many men were unable to serve in the military owing to poor health linked to inadequate nutrition. As a result, federal legislation in the United States was enacted in 1933 that provided loans to municipalities to pay for the preparation and service of lunches in schools. It was becoming clear that the health of the nation was directly tied to the health of the young. Two years later, the government began donating surplus farm commodities to schools, a practice that still exists (but to a much lesser extent). Thus, the noon meal soon became the standard in schools.

Later legislation, such as the National School Act of 1946 and the Child Nutrition Act of 1966, firmly institutionalized meals in school. While reimbursement rates have changed, largely due to the Omnibus Reconciliation Act of 1980 and related amendments in the early 1980s, school foodservice programs continue to make an effective contribution to children's nutrition.

The evolution of foodservice in colleges and universities is somewhat different from that found in primary and secondary schools. In the later Middle Ages, it was common for universities in Western Europe to include meals as part of student housing. Students ate and slept in the large halls that also served as study areas. This socialization process was as integral to student life then as it is today.

This practice also existed in colleges and universities in the United States as early as the 1800s. Later in the century, there was a move toward living outside of the university. With this trend came the growth of fraternities and sororities where students found opportunities to socialize, as well as the necessary housing and meals. Nonetheless, the convention of housing and eating on campus endured.

In the early 1900s, many universities featured dining halls where students gathered for a formal meal. This practice not only facilitated the socialization described earlier, but also offered the added benefit of teaching proper etiquette. This formal dining epitomized the stereotypical Ivy League image many held regarding college life, with formal dinners attended by young men in blazers and ties.

With the end of World War II came a large influx of college students. The new era was marked by the elimination of gender-specific dining halls and a move away from formalized dining. This corresponded to students' preference for more rapid meal service in concert with more choices. The new style of service allowed for greater numbers of menu items and for more flexibility in offerings, as illustrated in Figure 1.5.

In the 1990s, students in many countries became accustomed to an unprecedented array of menu choices coupled with customizable meal plans. Students could choose traditional **board plans**—which allow students to buy meals, usually served on an all-you-care-to-eat basis, in advance for the entire term—or opt for arrangements more akin to debit plans (where each purchase is automatically deducted from a prepaid account). Furthermore, students' expectations for familiar food sparked the large trend toward branding; most college and university foodservice featured more national brand products, such as Pizza Hut pizzas or Taco Bell tacos served from kiosks.

The expanded food offerings, which transcend the traditional ethnic-, region-, and religion-specific offerings of yesteryear, are also paired with nontraditional serv-

FIGURE 1.5 A Serving Line at Cornell University, Circa 1947. Copyright by Division of Rare and Manuscript Collections, Cornell University Library.

ing times. Matching the sometimes nontraditional eating habits of college students, most campus foodservice operations extend food offerings to multiple dayparts, beginning early in the morning and serving past midnight. Today, there is little that cannot be found in terms of menu choices and only very limited times when students cannot eat on campus in most large universities (see Figure 1.6).

Correctional Facilities

The idea of imprisonment, and the need to feed prisoners, is relatively modern. Until the mid-1700s, prisons served primarily to confine debtors who could not pay, those accused of crimes who were awaiting trial, or those waiting to be either transported or executed. At that time, prisoners were treated similarly to animals. They were given scraps and water—just enough to sustain life. In many cases, the population responsible for maintaining the prisons had little to eat themselves and little empathy for those behind bars. It was not uncommon for prisoners to survive only if their families brought food to them.

FIGURE 1.6   A Current Dining Outlet at Cornell University. Copyright by Cornell University.

Since the late eighteenth century, largely corresponding to a decline in capital punishment, prisons—more commonly referred to today as "correctional facilities"—have become instruments of punishment. This concept of the penitentiary originated in England and was introduced in the United States in Pennsylvania first and then in New York. The evolution of the modern prison was also due in part to the condition of the jails of the period, depicted so inaccurately in most movies set in the "Old West." In reality, the sanitation and care of prisoners in eighteenth-century English prisons, and to a lesser extent in American prisons of the same era and continuing into the nineteenth century, was so poor that outbreaks of disease commonly killed large numbers of the inmate population. This was at a time when there was no segregation according to gender, offense, or status (such as those pending trial vs. those awaiting execution).

While each country must deal with the nuances of its unique judicial system, all share the challenge of feeding inmates, regardless of how long a prisoner is in the system. The importance of this is underscored by the bloody 1971 Attica Prison riot in New York, which resulted in the deaths of 43 inmates and guards. One of the leading causes fueling the uprising was poor food quality.

In 1977 the American Correctional Association developed standards for **foodservice in correctional facilities.** Since then, many other agencies have adopted these standards. The stringent specifications focus on nutritional guidelines as well as quality and portion control.

Following a period in which many prisons were privately operated, most correctional facilities in the United States, at least during the twentieth century, were operated by the federal government in the case of federal penitentiaries or by state or local municipalities. This trend is now changing, with privatization becoming more common. The challenge remains, however, for each governmental or private agency: feeding inmates in a responsible yet cost-effective manner.

Recreational Facilities

Fans watching a game at a major-league stadium were once content with hot dogs, peanuts, soda, and beer. Today, every major stadium in the United States features a variety of food options, including but not limited to sushi, portabello-mushroom burgers, smoked-duck pizza, and fruit smoothies infused with invigorating herbs. The wider menu choices can also be seen at minor-league stadiums.

For example, the new PNC Park in Pittsburgh, Pennsylvania (see Figure 1.7), serves some 10,000 hot dogs during a game. But the on-site operators also serve hundreds of buckets of chicken wings, turkey focaccia sandwiches, hot corned-beef sandwiches (carved to order and served on fresh rye bread), chicken Caesar salads, and chicken fajita wraps. For younger fans, they operate a kids-only concession stand with no items priced over $2. And if that isn't enough, the stadium also features an Outback Steakhouse Restaurant!

The other facet of sports facilities that has evolved in the past decade is individual foodservice in sport boxes. New stadiums have an unprecedented number of these

FIGURE 1.7   Lunch at One of PNC Stadium's Suites. Copyright by ARAMARK Corporation © 2001.

FIGURE 1.8 European Sculpture Court Café, Metropolitan Museum of Art, New York City. Copyright by the Metropolitan Museum of Art.

private seating areas, sometimes cordoned off or separated by glass. Most feature waiter service, with a full menu and prices associated with careful food preparation and high-quality products. Whether at a racetrack or a football, baseball, or mixed-use stadium, **sports and recreation foodservice** is a key aspect of a fan's experience.

Have you been to a national park lately? Things have changed. The burger shack, which became commonplace in the 1950s, has been replaced by restaurants covering most types of foodservice, from quick service to fine dining. And here's a food order most experts would not have predicted even ten years ago: "Hi, I'm in campsite #24 with a green and blue tent; we'd like to order a large pepperoni and green pepper pizza with extra garlic." With the preponderance of mobile phones, many parks offer delivery among their foodservice options.

Museums, too, are now offering gastronomic delights to complement the art on display. Some on-site operations include multiple outlets, including grab-and-go type eateries and upscale restaurants with lunch menus featuring haute cuisine. For example, operations at the Metropolitan Museum of Art in New York City include a quick-service-style café, a large cafeteria, two lounges, a trustees dining room, a full-service restaurant, and a European-style court café, which features tables amid the various art displays (see Figure 1.8).

Convention centers and other recreational facilities, such as large ski areas, are also becoming more robust in their menu offerings. Whether at a boathouse on a lake or in a breakout room in a large conference center, most customers have become accustomed to food and service that rival those of traditional commercial restaurants.

Childcare and Eldercare Centers

Foodservice in childcare and eldercare centers is a relatively new phenomenon, one with a short history but a promising future. Largely the result of the growth of single-parent and dual-income families, childcare centers are becoming a niche business. The expertise required for operating childcare centers rests largely in the areas of education and child psychology. The foodservice component is also critical, however, since one of the basic tenets of providing care is proper nutrition.

Along the same lines, eldercare centers offer care for older adults who need help during the day when the primary caregiver, such as a child or sibling, is at work. The delivery of food in such situations is sometimes complicated owing to multiple dietary restrictions, the need to help with feeding, and the importance of delivering tasty, nutritious food. While the evolution of childcare and eldercare centers is still in its early stages, the need for foodservice acumen in this area is likely to increase as the market for such services continues to grow.

CHALLENGES OF ON-SITE FOODSERVICE MANAGEMENT

As each segment of on-site foodservice has evolved, so have today's customers. In 1955, 25 percent of the average American's food dollar was spent at restaurants. In 2000, this number had jumped to 46 percent. And this growth continues today.

Consider, too, these statistics from a recent National Restaurant Association study: Almost two out of five adults agree that using restaurants allows them to be more productive; more than half of all adults (53 percent) agree that the convenience of restaurants is critical to them.[6]

Changes in our society during the last several decades have produced customers with extremely high expectations of foodservice operators. Underscoring this, Phyllis Richman, the *Washington Post*'s food critic, recently said, "Restaurants are one of the primary ways we fill our bodies, occupy our social lives, spend our money, learn about the world, and conduct our business; they're the most central commercial arena in which we all participate."[7]

With the prominent position of foodservice in contemporary institutions, the trend toward greater customer expectations, then, is not surprising. Nonetheless, dealing with these savvy customers is challenging. And no segment of on-site foodservice is immune. In healthcare, for example, patients are simply not willing to endure tasteless food. Even when placed on restricted diets, such as low-sodium or low-fat diets, they expect food with flavor and menu items that reflect creativity. The other side of the two-edged sword—one that makes healthcare foodservice particularly challenging—is that cost-conscious administrators require meals that fall within stringent budgetary parameters. "Doing more with less" was never truer.

In B&I, cost and quality are important but in slightly different ways. Residents of office complexes are willing to patronize the in-house eatery because of its convenience. But the tether that keeps them in the building is extremely fragile. They may opt for any number of nearby chain-operated or local favorites. Thus, prices must be competitive, concepts must be fresh and reflect innovation, and quality must be consistent. The adage "You're only as good as your last meal" typifies today's B&I foodservice arena. In addition, B&I foodservice operators must combat the dissatisfaction of customers who, not long ago, enjoyed subsidized prices. Hence, operators must ensure that the perceived value of the menu items matches that of discounted items previously offered.

Such foodservice acumen among today's customers includes all age groups. Today's children eat out more often than those in any previous generation. They enter school with decisive food preferences, the satisfaction of which they have been conditioned to expect. Thus, they will be unlikely to settle for the stereotypical mystery meat in the cafeteria or stale snacks in vending machines. Rather, they demand foodservice that combines the glitz of leading quick-service restaurants (QSR) with the variety common in family-style restaurants. Foodservice operators, of course, must try to exceed these expectations at a minimal cost per meal. Moreover, they must contend with regulatory and parental guidelines for nutritional content.

In colleges and universities, freshmen arrive with educated palates and a keen orientation to good service. To meet the new demands, the menu today is as varied as most student populations; many campus foodservice operations feature food prepared to meet very specific dietary preferences such as vegetarian and vegan, macrobiotic, gluten-free, sugar-free, dairy-free, and newer trends like eco-cuisine. While some might argue that college-age individuals will always gravitate toward the nontraditional, whether in politics, religion, or food, this trend is more likely a direct response to the vast offerings of restaurants, where one can find hot dogs in Paris, sushi in Texas, and macaroni and cheese in Beijing.

Even prison foodservice reflects the new age of customer preferences. As mentioned earlier, the role of food in correctional facilities is very important from both a customer satisfaction and a nutritional standpoint. The latest trend toward the privatization of prisons has placed new emphasis on this, as well as on the cost component. Add to this the huge number of people—more than 1.6 million—incarcerated today, along with estimates of inmate population increases approaching 10 percent annually, and the need for extraordinary foodservice management replete with stringent cost control becomes evident[8] (see Figure 1.9).

Customers' worldliness in terms of foodservice is perhaps most noticeable in recreational facilities. As mentioned, the choices at ballparks dramatically outnumber the traditional offerings. This segment is unique, too, in the prices that customers routinely pay. The typical stadium patron doesn't blink an eye at paying $5 or $6 for a domestic draft beer in a paper cup. Surprisingly, many customers think they are being gouged simply because they form a captive marketplace. Some are coming to understand, however, that such prices are required to subsidize players' multi-million-dollar salaries. Unfortunately, stadium and arena foodservice operators—who achieve the same thin profit margins found in every other sector of foodservice—must also face patrons with expectations equivalent to the prices they are paying.

FIGURE 1.9 Inmates at Oneida State Prison in New York Package Bags of Beef Vegetable Soup. Copyright by State of New York—Department of Correctional Facilities.

OUTSOURCED VERSUS SELF-OPERATED FOODSERVICE

The original formula for a successful on-site foodservice operation required only a good manager to staff the outlet, design the menu, orchestrate the food production and delivery, manage the financial aspects, and—ultimately—serve the customers. As with everything else, over time the equation remained largely the same but the components grew in complexity, which was true in the foodservice department and in the **host organization** it supported. To meet the demands of this changing business, pioneers in the industry established systems to streamline operations, eventually offering their services to companies that needed on-site foodservice expertise. The selling strategy is fairly intuitive: Such entities as Kodak or Microsoft need to focus on their core competencies, which have nothing to do with managing foodservice. Thus, why not outsource such service to experts who can manage the enterprise more cost-effectively and with higher quality (owing to the systematized approach)?

This was the dawn of **managed-services companies** (also known as "contract-management companies" or "contract feeders"). Managed-services companies manage food and related services in a multiunit environment with reliance on a trade name; they depend on brand management as a vehicle for growth, and differentiate themselves on the basis of proprietary systems and approaches tailored to those on-site segments they service. The similarities extend to the human-resources component in that their people are their primary asset. In fact, managed-services companies differ from their chain brethren in that they have little in terms of physical assets; their people are, ultimately, what they bring to the table.

Using modern business models, the entire notion of outsourcing stems from a resource-based view of organizations. The argument often presented by resource-based theorists is that when a service or an organizational function does not deliver measurable value (such as profit) but is necessary to the operation, it should be outsourced to reduce the drain on resources better allocated to profit-producing endeavors.[9] This makes sense. If someone else can feed a company's employees cheaper and with better quality than the company can, alternatives must be explored.

So, are managed-services companies the panacea for the challenges of on-site foodservice? The answer, as one might expect, is "It depends." For some organizations, outsourcing is the perfect solution. The managed-services company assumes responsibility for everything, including the delivery of all food-related services. This frees the **client-based organization** (the company that is outsourcing) to focus on its core competencies.

In addition, most managed-services companies will tailor the relationship to fit the needs and specifications of the client-based organization, regardless of the circumstances. For example, in most B&I settings, the managed-services provider takes over the eatery, develops a menu and style of service that pairs well with the client-based organization, staffs the operations, and, in some cases, delivers a share of the profits to the client. In a different setting, the company looking to outsource the foodservice may want to keep the hourly foodservice employees on its payroll. (It may reason that it already has the infrastructure to support the employee base; the company may also want to manage the perception of other employees that it "protects" its employees even when outsourcing.) In such cases, the managed-services company will furnish the management team and will marry into the culture of the organization.

In the earliest applications, a client-based organization typically paid a fee to outsource its foodservice. In turn, the contracted company endeavored to save money through integration of its production systems, its labor management systems, and, of greatest importance, its purchasing ability. After all, the larger the managed-services company is, the greater the economies of scale are that it can realize through purchasing programs with vendors. The concept was that, once outsourced, the foodservice operation would be more cost-efficient (including the fee paid to the contractor) and produce a higher quality of foodservice.[10]

This fee-based relationship was very conducive to subsidized operations, since the goal of the operation was not to be profitable but to serve as a benefit to the employees. Managed-services companies were sometimes brought in even if the resulting cost was higher so long as the quality improved. In fee-based situations, the client-based organization commonly reimbursed the contractor for all expenditures in addition to paying the requisite fee. In the event that cash was collected, the managed-services company retained it and the corresponding amount was subtracted from the monthly invoice or the cash was given directly to the client-based organization. This was typical in both B&I and healthcare. In education, the fee-based structure was also common in the early years. Again, the intent of the client-based organization was to enhance quality by utilizing experts to manage food-related operations.

As business shifted to a more short-term focus on profit, beginning particularly in the 1980s, client-based organizations still wanted quality in their foodservice operations but also wanted a guarantee that the outsourced service would not exceed budgetary projections. This brought about two major changes in the contractual relationship. The first retained the fee concept but added a guarantee to the arrangement. Thus, a managed-services company would agree to run the eatery for a fee but would guarantee that all costs would not exceed certain limits based on the level of business. In some cases, contracts were written that provided an incentive to the managed-services provider for beating the budgetary numbers. In the event that the contractor exceeded the budget, such excesses were subtracted from the fee. In the unlikely event that the overruns exceeded the fee amount, the foodservice provider reimbursed the client.

The second change was a complete shift away from a fee mentality to one that directly linked risk and reward. In these contracts, commonly called "P&L" (profit and loss) formats, the managed-services provider is responsible for all financial aspects of the operation to such an extent that if a profit results, the contractor reaps the rewards. If a loss results, though, the contractor takes sole responsibility. This is more difficult than it appears since menu prices, staffing levels, hours of service, and other operational aspects are options specified in the contract. As a result, the managed-services provider must operate extremely efficiently in order to make a profit.

P&L arrangements are commonly found in B&I locations since the operations are fairly straightforward and clearly defined. Similarly, education can be structured under a P&L umbrella, although the contracts are more complex owing to the greater number of constituents. In healthcare, as well as in other segments, P&L contracts are becoming more common.

Clients considering outsourcing, particularly under a P&L format, almost always return to the issue of quality. After all, the managed-services provider could reduce the quality of products in order to bolster the bottom line—at least in the short term. This question is often answered through quality guarantees written into the contract between the two parties. In healthcare, for example, quality might be defined by third-party surveys sent to patients. The contract can specify certain parameters and financial implications for exceeding or missing the quality targets. Similarly, in education and correctional facilities, quality can be monitored through feedback from the primary users.

The other side of the coin pertains to "self-op" or self-operated foodservice outlets. While it is hard to argue with the potential economies of scale that large contractors offer, organizations may also hire individuals to manage their foodservice operations just as they hire managers for any of their other business units. In the case of large organizations, such as healthcare providers involving dozens of hospitals, building an infrastructure and internally developing the talent to run multiple foodservices can be attractive. Even in small operations, such as a small assisted-living center, the opportunity to operate without outsourcing is a viable option. The decision rests largely on a number of factors, including the history of the operation, the culture of the organization, the talent level of the current foodservice management team, and the ideological stance of the company's board.

So how many foodservice operations are outsourced? The exact number is difficult to pinpoint. In the United States, experts estimate that approximately 50 percent of noncommercial foodservice is operated by managed-services companies. While this is a rough approximation, the numbers become somewhat more precise when broken down by segment.

As shown in Figure 1.10, B&I is dominated by managed-services companies. This reflects the rampant downsizing of the past 15 years and the trend toward outsourcing noncritical functions within organizations (not to say that foodservice is not important). Aside from those focusing exclusively on healthcare, almost all managed-services companies have a presence in B&I. Colleges and universities are also popular owing to the magnitude of the operations and the associated economies. Furthermore, the recent trend in university foodservice to include national brands makes it a natural fit for leading managed-services companies that have multiple franchise agreements with such chains as Burger King, Taco Bell, and Subway. Managed-services companies' presence in hospitals is a natural response to the cost focus of most institutions. Also, the widespread capitation of costs in healthcare reimbursements underscores the need for experts in every area, particularly support services such as foodservice, where the opportunity to save money is attractive.

Foodservice in recreational facilities, which has grown more than any other on-site segment in terms of annual revenues, is becoming increasingly competitive for managed-services companies. However, this is a market only for the larger managed-services companies; most sports and leisure facilities require a considerable cash investment on the part of the contractor in exchange for a long-term agreement. This investment is used to enhance foodservice-related facilities, thereby creating greater opportunities for increased customer service and revenue enhancement.

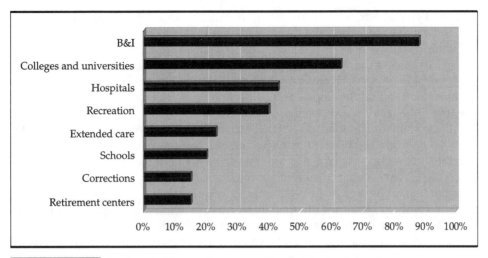

FIGURE 1.10 Managed-Services Companies: Penetration by Segment.

Contracted foodservice in nursing homes and other extended-care subsegments of healthcare, including retirement communities, is growing, albeit slowly. Many smaller organizations already have optimized foodservice; in addition, managed-services companies often cannot generate enough profit from smaller operations to justify the corporate support required to maintain systems and, in turn, quality. Smaller schools are in a similar situation. Managed-services companies are now, however, managing foodservice in entire school districts in many parts of the United States. State and federal cost-reduction programs and the cost savings possible through hiring efficient foodservice experts have increased the demand for managed-services companies.

The stringent specifications of correctional facilities are a good match for the systematized approach of a quality managed-services company. Correctional facilities have often struggled to find quality managers for support services owing to the inherently challenging situation. Managed-services companies relieve this burden. Most contracts are also based on a per-meal or per-inmate guaranteed cost, making outsourcing particularly attractive as many prisons seek to reduce expenses.

Childcare and eldercare centers are looking more toward managed-services companies that can provide meals delivered from a separate facility. Thus, a managed-services company with a contract with the local school district is an attractive partner. Other options are managed-services companies that prepare meals at tertiary locations in the same area. Such operations are more common in larger urban settings.

LEADERS IN ON-SITE FOODSERVICE

Whether you are against the concept of outsourcing foodservice, possibly arguing that it results in loss of control, or you believe outsourcing is the only alternative in this era of rapidly developing business, it is hard to ignore the changes in the managed-services arena. For example, two articles, one written in 1997 and the other in 1999, were intended to describe the industry leaders at the time. In just those two years, the difference was astounding. The four largest companies became three, owing to a merger. And all of the firms grew substantially due to both organic growth and acquisitions. Moreover, each company had continued to pour unprecedented amounts of money into brand development, new technologies, human-resource information systems, and management development.[11]

These changes continue today. Even textbooks that devote entire chapters to describing the scope and reach of managed-services companies are outdated before the ink dries on the pages. Thus, it is more practical to segment the market in terms of basic organizational characteristics and coverage. Further analyses and descriptive information by company name are better reported annually by trade publications.

Early authors segmented the managed-services arena according to a tiered system whereby each tier reflected the size, regional coverage, sales, and number of contracts. A better approach is simply to categorize companies into two groups based largely on size. The first group comprises numerous small and midsized companies that generally target specific geographic regions, are organizationally centralized, focus on a limited number of market segments, and concentrate on a single core service—foodservice versus, say, a combination of housekeeping and facilities-maintenance services. Annual revenues vary considerably but are generally less than $250 million. A good example of such a company is American Food and Vending (see the company profile).

The second group comprises the biggest players in the global marketplace, with substantial penetration in several countries. Maximizing the economies resulting from sales in the billions, these firms capitalize on their global experience and collective managerial acumen. Each firm operates in a number of segments, and most are highly diversified in the managed services they offer. With the lion's share of the market in several countries, these firms are most notable for exhibiting all the characteristics of major-brand chain restaurants.

These companies foster their key asset—their people—like no others and spend the development dollars to prove it. For example, the North American division of the on-site giant Compass Group PLC, headquartered in the United Kingdom, received awards in 1999 from The Communicator, an organization that recognizes excellence in corporate video programs, for their training videos. Further emphasizing the internal management of the global leaders in managed services, Sodexho Alliance, another global giant—this one based in France—recently celebrated its inclusion in *Fortune* magazine's list of the world's most admired companies. Finally, ARAMARK, the only global leader with its primary corporate office in the United States, demonstrates its commitment to its people, constantly emphasizing the role of its location managers through both its marketing and its daily business practice. (For more information on ARAMARK, see the company profile.)

Regardless of their size, scope, reach, position, or home country, managed-services companies are an integral part of the foodservice industry. While they are certainly not the only source of the best practices, they continue to push the envelope in terms of systems development and technological innovations.

American Food and Vending—A Regional Company in Transition

Deliver Best Value to Our Clients and Guests at Every Moment of Service is the vision that has driven American Food and Vending since Martin Wells founded the regional on-site foodservice provider in 1990. Joined by his brother Steve, also an accomplished attorney, Martin has drawn from common experiences the Wellses shared in growing up in the on-site foodservice business.

From the outset, American's leaders focused their business strategy exclusively on the B&I segment, encompassing both dining and vending services. They observed that the country had many small vending companies that, while capable of providing vending services, did not have the expertise to provide fine dining foodservice or the capital to compete with national managed-services companies. They perceived, however, that they could differentiate themselves from the larger companies by staying close to the customer, an option made possible through their regional positioning. In addition, the company operates on the premise that vending and dining are not separate entities.

Does this business model work? As of 2002, the company has experienced 25 percent annual growth since its inception.

American sees the ability to maintain minimal overhead costs through a focused infrastructure within a competitive environment as a key competitive advantage. Simply put, the company can operate at lower profit levels per site while retaining sufficient earnings to fuel expansion and meet or exceed service expectations.

Another successful approach is operating contracts on a P&L basis. This shifts the risk of loss from the client to the operator. American has gained a significant amount of business by utilizing its cost-saving expertise to help numerous clients reduce operating expenses previously allocated to foodservices.

In concert with these competitive advantages, American has a number of programs that contribute to its growth and stature as a regional contractor. For example, the "Amerimart" concept mimics the local convenience store and provides soups, hot and cold sandwiches, pizza, and a handful of hot entrées. A range of capital investment options is offered based on client populations. A key to successful implementation is efficient space utilization and service format, which significantly reduce operating and labor expenditures.

American operates both large and small on-site operations. In meeting potential clients, American's employees quickly point to a number of reasons that American is a good partner in business:

- The organizational structure facilitates immediate decisions and a quicker response to service requests.
- Because American provides 100 percent focus and expertise in both food and vending services, clients have the option of using one vendor for both services, a more convenient and cost-effective way to operate.
- Private ownership and the single-segment focus prevent internal competition for resources. This allows energy to be directed at exceeding customer expectations.
- American's owners and key team members have a vested interest in the company and in most cases are long-term employees. This supports business retention through lasting relationships with clients.
- A cookie-cutter approach is never taken. Menu, service style, and hours of operation are designed to meet the specific requirements of each client corporation.

Now operating in eight states, American has emerged as a leading regional company. The corporate vision is to retain this position and manage growth intelligently through planned expansion into other key markets. Throughout, the Wellses have realized that the business is complex yet simple. When the dust settles, it boils down to a matter of exceeding client expectations in every situation—at every moment of service.

COMPANY PROFILE
ARAMARK Corporation

ARAMARK (stock exchange symbol: RMK) is a leading provider of a broad range of outsourced services to business, educational, healthcare, and governmental institutions and sports, entertainment, and recreational facilities around the globe. The company currently is a market leader in food and support services, uniform and career apparel services, childcare, and early education. Depending on the year and activity in the marketplace, ARAMARK is one of the three largest foodservice companies in the United States. In most of the other countries in which it operates, the company is also among the top three.

Like most businesses, ARAMARK started small. In this case, the inventory was peanuts, the warehouse was the back seat of a 1936 Dodge, and the man with the big idea was Davre Davidson. His vision was to put vending machines inside factories and offices—places they had never been before. In his mind, his business was not vending, it was service, a concept shared by another like-minded entrepreneur, William Fishman.

The same customer, Douglas Aircraft, brought the two together during World War II. They shared ideas, became fast friends, and at war's end began offering foodservice as well as vending. In September 1959, Davidson and Fishman merged their operations into one under the name Automatic Retailers of America, or ARA for short.

Davidson's operations were primarily on the West Coast, Fishman's in the Midwest. To be truly national required an East Coast presence as well.

It came in 1961, with the acquisition of the Philadelphia-based Slater Company, the country's largest foodservice business at the time. Vending and foodservice made ARA a diversified service provider. It would continue to diversify over the next several decades.

ARAMARK's continuing goal is to capitalize on favorable outsourcing trends by offering a large and diverse client base an expanding portfolio of services to meet their outsourcing needs. In fiscal 2000, the organization reported sales of approximately $7.3 billion and net income of approximately $168 million. Over the past five years, primarily by expanding food and support service offerings, maintaining a diverse base of existing client relationships, maintaining high retention rates, and increasing uniform capabilities, ARAMARK has achieved compound annual sales growth of 8.1 percent and compound annual operating income growth of 12.8 percent, adjusted to exclude two divested noncore business lines. The executive team believes they will continue to grow the business rapidly by capitalizing on the continuing growth of the outsourcing market, the company's market leadership positions, and the added access to capital that the recent public offering provided.

For over half a century, the company has emphasized the uniqueness of each client relationship with a whatever-it-takes attitude. For ARAMARK's employees, who number more than 100,000, it is simply an acknowledgment that in true partnerships, there are no conditions.

CHAPTER SUMMARY

On-site foodservice includes food outlets in B&I, schools, universities and colleges, hospitals, skilled-nursing centers, eldercare centers, correctional facilities, recreational facilities and childcare centers. Sales in these outlets represent more than a quarter of worldwide foodservice expenditures. In terms of market share, B&I and educational foodservice represents over half of the total revenues from on-site foodservice, followed by sales in healthcare settings.

The history of each segment is closely tied to the related business it serves. Healthcare foodservice, for example, gained complexity as the medical profession fully evolved into its present form. Similarly, B&I foodservice came into its own during World War II, when competition for labor reached a peak and employers sought to attract employees and keep them close to work by offering meals. School foodservice, too, evolved from an initial attempt to use food to encourage attendance.

The challenges for on-site foodservice managers are largely linked to societal changes, such as a substantial increase in dining outside the home. These changes have resulted in a savvy customer base that expects quality and value whether dining in a hospital cafeteria, an office complex, or a national park. Even in schools, children accustomed to dining out frequently have mature tastes and staunch dietary preferences. In college, this trend is even more dramatic; customers expect menu items that reflect diverse ethnic flavors and unique dietary preferences.

The on-site arena is often divided into two categories: outsourced foodservice, operated by a managed-services company, or a self-op wherein the host organization operates the foodservice establishment internally. Both configurations have merit. Many companies choose to outsource their support services such as foodservice so that they can concentrate on their core businesses. Others prefer to retain complete control of all facets of the company and proudly tout their self-op culture.

Those companies that do outsource have a variety of firms from which to choose. These client-based organizations can choose from sources ranging from regional providers to multinational contractors, many of which offer multiple services. The contractual relationship between the managed-services company and the client-based organization can be fee-based—in which the contractor is simply paid a fee—or it can be based on a P&L format in which the contractor is responsible for all operational aspects and receives a profit only if sales exceed expenses. Today's contracts also include hybrids replete with quality guarantees.

KEY TERMS

on-site foodservice
healthcare foodservice
Code of Hammurabi
capitation
business and industry
    (B&I)
subsidized meals
school foodservice

board plan
correctional facilities
    foodservice
sports and recreation
    foodservice
childcare center
    foodservice

eldercare center
    foodservice
host organization
managed-services
    companies
client-based
    organization

1. Define on-site foodservice.
2. What commonalities do healthcare foodservice and B&I foodservice share based on the evolution of each segment?
3. Why is the Code of Hammurabi pertinent to the discussion of the evolution of foodservice in healthcare?
4. Why is capitation important in understanding the healthcare industry?
5. How important do you think subsidized meals are to employees in today's economy? What does this imply for employers?
6. How did foodservice change from the time you were in elementary school to your early college years? Were these changes positive?
7. From a foodservice perspective, how have correctional facilities changed over the years? What was the most important event in recent history that changed prison foodservice?
8. What advantages do managed-services companies offer to client-based organizations? What disadvantages can you imagine from the perspectives of both the client and the (former) foodservice director?
9. Referring to the Preface, why study best practices? What do these mean for foodservice managers?

END NOTES

[1] Reynolds, D. (1997). Managed-services companies. *Cornell Hotel and Restaurant Administration Quarterly, 37*(3), 88–95.

[2] For more information, see Hurowitz, V. A. (1994). *Inu Anum ṣīrum: Literary structures in the non-juridical sections of Codex Hammurabi.* Philadelphia: University Museum.

[3] Harris, A. T. (1967). *A textbook of hospital catering.* London: Barrie & Rockliff.

[4] For more on this, see Grandshaw, L., & Porter, R. (1989). *The hospital in history.* London: Routledge; and Long, D. E., & Golden, J. (1989). *The American general hospital.* Ithaca, NY: Cornell University Press.

[5] For more information, see Owen, R. (1972). *Owenism and the working class.* New York: Arno Press; and Claeys, G. (1987). *Machinery, money, and the millennium: From moral economy to socialism, 1815–1860.* Princeton, NJ: Princeton University Press.

[6] For more information and related statistics, see Editorial Staff (December 1999). Restaurant-industry real sales growth 1971–2000. *Restaurants USA, 19*(11), F2–F28; and National Restaurant Association & Deloitte & Touche, LLP (2001). *2001 Restaurant industry operations report.* Washington, DC: National Restaurant Association.

[7] Panitz, B. (February 1999). Unwrapping what the industry has to offer. *Restaurants USA, 19*(2), 26–30

[8] For more information, see Wagner, C. G. (1999). Population boom in prison. *The Futurist, 33*(10), 8; Sheridan, M. (2000). Corrections. *Restaurants & Institutions, 110*(25), 82–84; and Matsumoto, J. (2000). Serving time. *Restaurants & Institutions, 110*(3), 65–68.

[9] For seminal work in this area, see Penrose, E. (1959). *The theory and growth of the firm.* New York: John Wiley & Sons.

[10] For a rich depiction of how one of the legends in on-site foodservice began, see Scandling, W. F. (1994). *The saga of Saga: The life and death of an American dream.* Mill Valley, CA: Vista Linda Press.

[11] Reynolds, D. (1999). Managed-services companies: The new scorecard for on-site foodservice. *Cornell Hotel and Restaurant Administration Quarterly, 40*(3), 64–73.

CHAPTER 2

ORGANIZATIONAL STRUCTURE AND THE FOODSERVICE DEPARTMENT'S ROLE

The history of on-site foodservice in various settings discussed in the previous chapter provides an interesting foundation for understanding how the foodservice department fits into the organizational structure of the host organization (whether self-op or contracted) and the department's role within the organization. Traditionally, foodservice was but one support department, and its role was straightforward: to produce meals. Today, however, the organizational structure is more organic—the foodservice department works more synergistically within the organizational framework. Moreover, the department's role has changed, with duties that extend far beyond simply providing lunch to employees. Today's customers include employees, guests, external customers, and, in healthcare, both inpatients and outpatients—those receiving care on an outpatient basis. In addition, most outlets today serve multiple meal periods, known as **dayparts,** in response to the extended workday that is now commonplace. And the service doesn't stop there. Many foodservice departments offer home-meal replacements, such as roasted chicken dinners and whole pizzas packaged to go.

As most will attest, the business world is changing with the pace and undulations of a roller coaster. Not surprisingly, such shifts in business practices and mindsets have dramatically altered the foodservice department's role. This chapter first addresses the changing nature of the host organization, again using healthcare as an exemplar. Next, the foodservice department's response to such changes is illustrated. The chapter concludes by considering differences in how customers are regarded, as they are now viewed as more than simply the recipients of meals. Such changes illustrate how managers in every segment of on-site foodservice must evolve in order to remain at the frontier of delivering the best in food and service.

DOING BUSINESS IN TODAY'S CHANGING ECONOMY

Through the end of the twentieth century and into the early months of the twenty-first, most analysts agree, the economy of the United States and several other countries was booming. Following a slowdown in the early 1990s, most firms were bolstering bottom lines through increased sales and reduced expenses. In recent years, the economy has slowed, picked up a little, slowed a little. How healthy the economy seems depends today on who is asked to describe it.[1]

Whether one believes the economy is expanding or not, there can be little argument that business is different from the way it was even ten years ago. The focus is on cost containment and efficiency; this extends to travel budgets, advertising, and, of considerable importance, noncore businesses such as foodservice for employees. This is obvious in high-tech firms—which are renowned for unusually generous workplace incentives—that commonly offered perks ranging from free doughnuts and snacks for employees (under the argument that such an offering enhanced productivity) to subsidized meals. This is no longer the norm.

Motorola Inc., a global leader in providing integrated communications and embedded electronic solutions, provides a vivid example of the new approach to business. In 2001, the company saw revenues from foodservice in excess of $55 million, a result of cafeterias, catering, and vending operations in plants across the United States. The tertiary business was so large that it was operated as a separate division, called the Motorola Hospitality Group, and was led by a corporate director with responsibility for all hospitality-related services.

In the late 1990s, Motorola executives decided to challenge the division to justify its existence within the firm—after all, Motorola is known for outsourcing anything it does if someone else can do it better and cheaper—by pitting it against several managed-services companies for the right to continue operating Motorola's cafeterias. The in-house group won the bid, and in the process outlined a plan for reducing operating expenses by $6 million (a goal the division soon achieved) while also promising to complete various renovation projects (see Figure 2.1).

At the turn of the twenty-first century, Motorola Hospitality Group was the largest self-operated enterprise in America. The battle to drive up revenues, reduce costs, and enhance quality does not end, however. In the latter part of 2001, Motorola Inc. hired a global managed-services company to run its foodservice operations. Under the agreement, the contractor agreed to retain all of the division's employees and management personnel. In exchange, the contractor assumed operations of all Motorola foodservice units including those in Illinois, Arizona, Texas, and Florida.

Challenges in the Healthcare Industry

Healthcare providers felt this shift before most companies in other industries owing to rampant changes in the industry that started as early as the 1980s, when a largely unforeseen confluence of forces occurred in the United States. First, the cost of associated goods and services required by healthcare providers increased in excess of general inflation. Next, expenditures for Medicare and Medicaid increased dramatically (from $6 billion in 1967 to more than $72 billion in 1981), resulting in governmental pressures to contain healthcare costs.

This was when the Reagan administration created its **diagnosis-related group** (DRG) payment system in 1982. The DRG system is based on prospectively determined prices or rates rather than retrospectively determined costs. All patient diag-

FIGURE 2.1 The Newest Motorola Employee Cafeteria. Copyright by Motorola Hospitality Group.

noses are grouped into hundreds of DRGs, each having a fixed payment. Hospitals are paid these fixed fees regardless of the costs incurred for treating patients.

During this time, many hospital administrators saw an opportunity to combat such fiscally challenging situational conditions through mergers. The potential synergistic effects were impressive: Merged healthcare providers realized increased market power, economies of scale, and the availability of cash generated from increased services. As a result of early gains, the number of mergers and acquisitions reached over 200 per year in the early 1980s. This was a dramatic jump for an industry that had traditionally avoided such strategies. (For example, an American Hospital Association survey found that only five hospitals had entered into any type of combination with another hospital in 1961.)

At the same time, **health-maintenance organizations** (HMOs), which are healthcare plans that provide comprehensive medical services for employees and their families at a flat rate, were springing up and offering insurers the opportunity to reduce the cost of healthcare dramatically. This was tremendously attractive to employers who had traditionally paid employees' healthcare costs, usually on a fee-for-service basis, and were revolting against the ever-rising healthcare costs.

The cost savings, however, were realized through capitated reimbursement plans, which resulted in lower revenue for healthcare providers. (As noted in Chapter 1, the model underlying capitation features an HMO that makes a set monthly payment for each enrollee. That payment is expected to cover any and all healthcare services provided by a hospital, physicians, and ancillary services.) In a nutshell, the ascendancy of HMOs, while beneficial in that they were originally intended to reduce ultimate costs by fostering preventive healthcare, resulted in lower utilization rates for hospitals (owing to a host of factors). This reduction necessitated unprecedented cost cutting. Hospital administrators—whether in for-profit or not-for-profit settings—began scrutinizing every facet of their organizations as never before, analyzing costs pertaining to new surgical equipment as well as paper supplies in the cafeteria. They also began to consider how revenue could be enhanced given the challenges created by preset reimbursement rates.

By the late 1980s and early 1990s, healthcare costs remained out of control, with increases commonly in the double digits. HMOs were encountering their own crises, brought about by the challenges associated with huge growth and the reality of a public that demanded state-of-the-art medical care. Whether the reimbursement for such procedures was capitated by HMOs was immaterial to the end user.

With increases in operating expenses returning to more modest levels—but still far exceeding inflation—healthcare costs in the United States became a national concern in the late 1990s and remain so today. Some believe HMOs are the problem, while others contend that some hybrid of the original model holds the answer. Those who are most familiar with the industry posit that a true solution must include participatory reengineering by all constituents involved in the healthcare environment, including consumers, physicians, administrators, insurers, and suppliers. One thing is clear: The traditional business approach for hospitals has gone the way of the dinosaurs. Today, healthcare providers must operate in a customer-driven economy and must learn to manage costs in order to survive. Furthermore, the emphasis is on revenue maximization, with the goal of adding only those expenses that add a disproportionately greater increase to the bottom line.

THE EFFECTS ON FOODSERVICE OPERATIONS

The healthcare industry is not unique in its quest to control costs and increase efficiency, as was noted in the earlier example of Motorola. The tremendous challenges it faces do, however, underscore the unprecedented approach to cost cutting that so many industries are only recently encountering. And as witnessed across industries, administrators look first to cut costs in nonprimary areas, such as tertiary services.

One might think this is an approach that yields limited returns. In reality, such an approach can be quite effective in the short term if the size of the support service is substantial. Such is the case in large healthcare institutions, where foodservice represents a multi-million-dollar enterprise. Consider an urban healthcare provider such as the Boston Medical Center, a 547-licensed-bed, academic medical center with more than 3,300 **full-time equivalent** (FTE) employees. (An FTE represents a 40-

hour-per-week position. It may be filled with two or more part-time employees, but the hours paid for the position remain at 2,080 per year.) Such an institution must feed (1) the patients occupying the rooms, which alone represents more than 450,000 meals a year; (2) the outpatients, which, assuming each has a single meal, could represent another 400,00 meals; (3) visitors to patients and those accompanying other patients; and (4) employees who staff the hospital 24 hours per day, 365 days per year. In sum, it would not be unusual for such an institution to produce more than 2 million meals per year.

Thus, when an administrator must reduce costs, he or she will, logically, target a department that represents a large dollar figure on the expenses section of the income statement. This makes foodservice a prime target. At first, administrators typically look at the foodservice department budget and quickly identify labor as the largest expense. In the 1980s, it was not uncommon for foodservice managers to receive mandates for a 10 percent (or greater) reduction in FTEs. Some institutions, anticipating the need to cut costs, accomplished initial cuts through natural attrition, thereby avoiding the negative perception that jobs were being sacrificed. In the 1990s, however, staff reductions were severe and typically replaced the practice of reduction through attrition with reduction through pink slips.

The Role of Foodservice as Part of the Host Organization

This cost-cutting approach effectively reduced labor costs associated with the foodservice department, but it was a short-term fix. The foodservice positions commonly eliminated included sanitation workers, administrative assistants, and serving staff. While some gains were made through cross-training and reworking job flows (more is said about this in Chapters 8 and 9), these cuts had an inverse reciprocal effect on quality. Keeping the department clean and organized, closely monitoring expenses through data entry, and serving all categories of customers becomes more challenging with fewer persons to do the work. Everyone quickly realized (some more quickly than others) that the slash-and-burn approach was not the answer.

Administrators and managers recognized that the foodservice department could no longer be treated as an autonomous entity, one that could be analyzed only from the perspective of expense reports. Instead, the foodservice department needed the same attention that every other department received. In essence, foodservice is as much a part of patient care as many other functions. Returning to the holistic notion that food and nutrition are pathways to better health, foodservice managers began to think about their departments as an important function of the organization.

This realization had direct meaning for cost control. In the old view, food, labor, and direct expenses were just part of the process of producing food. Now, every line item on the financial statement was regarded in terms of how much value it contributed to the end user. Granted, hospital foodservice was always predicated on a certain level of cost consciousness, but this new outlook integrated service into the equation.

This fresh perspective meant looking at everything, justifying everything, and using zero-based budgeting. For example, certain food items are typically stocked in pantries near the nursing station in the event that a patient needs a snack. Since the goal is to provide adequate nutrition to patients, this practice makes sense. However, in many institutions, the pantry stock also served as a snack bar for employees. Analysis of patient consumption by room quickly brought the excesses to the forefront. Thus, policies were adopted that precluded employees from snacking on food intended for patient consumption, resulting in thousands of dollars in savings for some organizations.

Foodservice at the Ohara Group in Japan provides this chapter's first example of best practices owing to its ability to develop and provide services that fit the needs of an organization that covers a spectrum of healthcare-related services, including acute care, assorted outpatient services, rehabilitation services, assisted living, long-term care, and home healthcare.

FOODSERVICE: AN AGENT OF CHANGE AT THE OHARA GROUP (KYOTO, JAPAN)

The Ohara Group, known largely for the Ohara Memorial Hospital and Elderly Care Center in Kyoto, Japan, is a leader in healthcare in Japan. Once a small, stand-alone facility, the organization has evolved into a complete-care institution, dramatically expanding its services to meet the everchanging needs of its diverse market. Today, Ohara provides everything from inpatient surgery to home healthcare (see Figure 2.2).

For the foodservice department, this evolution was somewhat difficult. Originally, foodservice was responsible for producing meals for patients and staff in the same building that housed all the associated healthcare services. Those were simpler times, and an able staff using in-house kitchen facilities readily met the needs of the institution, producing fewer than 1,000 meals a day. With the addition of the Roken (a facility designed to provide care for patients convalescing for 60 to 90 days) and the Care House, the number of meals served more than tripled. To meet these needs, the foodservice department integrated cook-chill technol-

ogy (discussed at length in Chapter 3). Hence, they were able to retain the central kitchen and did not need to increase the production staff dramatically, as would have been necessary had they added satellite kitchens.

With the addition of daycare and outpatient services, the foodservice team expanded the production area only slightly to meet the need for some 300 additional meals a day. With the advent of home healthcare, however, the problem of producing and delivering approximately 800 meals a day was not so easily surmounted. In simple terms, there was no way the foodservice operation could produce so many additional meals without substantial capital improvements, more kitchen space, and a much larger staff. The productivity of the department was impressive, but no matter how the many professionals analyzed the situation, the foodservice system's capacity simply could not be expanded.

Other attributes of the new meals also deserved consideration. First, cold meals needed to

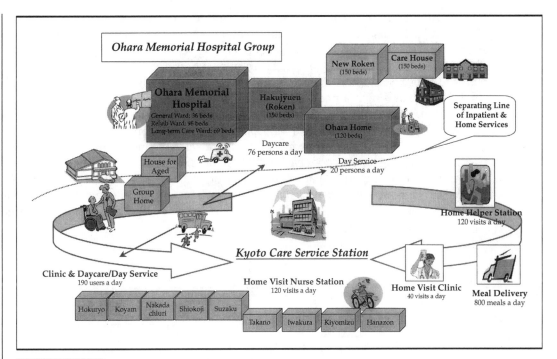

FIGURE 2.2 Ohara Group's Range of Services.

be held at the appropriate temperature. This would require considerable cold storage in addition to the space needed for staging the meal production. Second, hot meals—if delivered hot—would need to be held at the proper temperature for extended periods of time, potentially compromising food quality. Moreover, the prospect of foodborne illness was omnipresent whether the food was transported hot or cold. Finally, the logistics of meal delivery (in addition to the temperature issue) was outside the expertise of the current staff.

The answer? The savvy managers outsourced meal delivery to a company experienced in home-meal delivery. Thus, Ohara was able to maintain quality, which is in part responsible for their good reputation, without adding additional operating expenses to the department. Granted, the incremental cost of each meal served was higher, but the long-term benefit, including the reduction of risk, was more than worth it.

Today, Ohara serves close to 2 million meals a year through a combination of in-house and outsourced venues. They provide the best practice in leading rather than following the charge toward change. The foodservice management team didn't wait to see what would happen; they made changes in anticipation of the organization's many evolutionary phases. They have continued to maintain their reputation for both high-level healthcare and quality foodservice, and—of equal importance—the foodservice department is regarded as a vital, productive, and fiscally responsible part of the organization.

Another recent change is a new reporting relationship for the foodservice director. In past years, the foodservice director typically reported to an operating vice president, who had responsibility for other operating departments such as materials management or emergency services. There often was no rhyme or reason to how the foodservice department fit into the reporting structure of the institution. For example, Figure 2.3 illustrates a traditional hospital organizational framework.

Today, foodservice is often bundled with other hospitality-related services such as housekeeping, communication, and patient transportation. Directors from each of these areas report to the same individual, thereby facilitating interdepartmental communication. Again, this goes back to the notion that foodservice is not just a support service; it is an important part of the organization, supporting the overarching goal of providing complete, holistic service to every customer. The alignment of hospitality services, discussed at length in Chapter 13, also helps emphasize the customer rather than the organization.

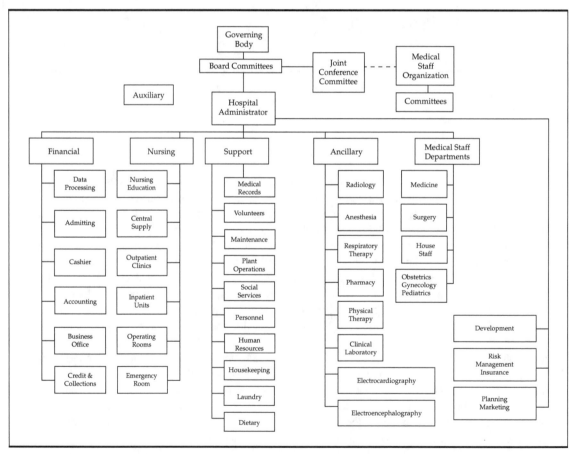

FIGURE 2.3 A Traditional Organizational Configuration for a Typical Hospital.

Foodservice Isn't Just a Hot Lunch Anymore

With hospital administrators focusing more on revenue, foodservice managers have begun to look at business in terms of a more traditional restaurant model. That is, revenue needs to exceed expenses if the entity is to survive. With the commensurate shift away from subsidized employee meals found in other business sectors, foodservice operators began analyzing the employee cafeteria in a new light. Traditionally, prices were set to reflect an apportioned subsidy. Guests who braved the hospital cafeteria were often asked to pay a somewhat higher price; this was generally done on an honor system on the part of the customer.

Savvy foodservice operators, however, realized that if revenue could be increased in the cafeteria, drawing from employee and guest patronage, then quality enhancements in the form of increased labor and higher quality (i.e., more expensive) menu items could be integrated on the patient side of the equation. This same approach was applied to vending and catering. Each aspect of the foodservice department, including staff dining and public feeding, patient services, and catering, along with best practices corresponding to each area, is discussed in detail in Chapters 4, 5, and 6, respectively.

REDEFINING THE "CUSTOMER"

Moving to a model in which the foodservice department provides a service to the organization involves quite a few changes, not the least of which is a new approach to managing the foodservice operation. For most, this is a learning process, one that is constant and involves ongoing discovery based on new situations and multiple sources of input.

Patient-Focused Care

One of the first steps in dealing with the foodservice department's new role in the organization requires a primary focus on the customer. In healthcare, this is commonly referred to as **patient-focused care.** Patient-focused care has become a central element in work redesign and restructuring for healthcare delivery systems of all sizes.

Through restructuring, reengineering, or redesigning, organizations in most industries are doing everything possible to get closer to the customer, not only to be the first to market but also to be the first to come to customers' minds. In this context, patient-focused care is nothing new. However, traditional reengineering, under whatever name, typically consists of reorganizing the way in which services or products are delivered through multiskilling or cross-skilling, including adapting old delivery channels and adopting new ones. Patient-focused care, on the other hand, involves substantial transformation of the organizational structure but also entails shifting the service provider's attitudes, which influence the way in which care is provided.

To facilitate the shift in culture from provider-driven to patient-focused care, top management must lead the charge. Even when driven from the top down, however, such organizationwide change fosters **creative tension,** which causes uneasiness between what is known and what is envisioned.[2] This is particularly problematic in foodservice operations, where many employees perceive change as negative or, even worse, as equating to a greater workload. Dealing with creative tension is worthy of a book in itself; some suitable approaches are highlighted in Chapter 10.

The patient-focused approach fosters a culture of caring and comparison that transcends departmental barriers. For example, employees in nursing and foodservice often find themselves at odds, blaming one another for patient dissatisfaction. Using the concept of patient-focused care, such blaming is supplanted by trying to find solutions. Both departments focus on the patient and on building a relationship with that patient. Comments such as "Why is my food cold?" can provide an impetus for problem solving instead of faultfinding. At least that is the goal!

Meeting Customers' Needs in New Ways

Look at almost any good on-site foodservice operation today and one thing is apparent: It is more than a food outlet. It is part educator, providing information on health-related topics, most often relating to food. This is evident in other segments, too. Quick-service restaurants include nutritional information for various menu items, and midscale restaurants often highlight low-fat options. It is also part entertainer. Menu boards are designed with artistic flair; dining areas are sometimes wistful, other times artistic, designed to induce a relaxing, even mood-altering, experience. In most settings, in fact, the intent is to dissociate the eatery from the host organization. Thus, color schemes in the dining areas may be in stark contrast to those found in nearby corridors, offices, and public spaces.

In addition, on-site foodservice is often considered an employee benefit as part of an effort to improve the health of the company's constituents. The advent of salad bars was an early indicator of this trend. Implemented partly to give customers greater variety, salad bars also appealed to those seeking healthier cuisine (the benefit of a plate of iceberg lettuce drenched in bleu cheese dressing aside). From these humble beginnings, on-site foodservice began a subtle segue to part food provider, part contributor to customer wellness.

This has resulted from employers who understand the link between better health and higher retention of their employees, as well as lessened healthcare costs. In other words, prophylactic efforts aimed at reducing medical costs associated with an aging workforce (as discussed earlier) equate to healthier bottom lines. It is simply good business.

This emerging role of foodservice providers as champions of customer health is also the result of concerns over **burnout.** Typically the result of overextension, burnout results in higher medical costs, higher turnover, and lower overall job satisfaction. Ironically, on-site foodservice, particularly in the B&I segment, was integrated to increase productivity. As discussed in Chapter 1, the goal was to keep employees in the building and to keep the focus on work. Now the on-site foodservice operations in most corporate offices are part of a plan to allow employees to put work out of their minds, at least during the meal or break period. The new goal is to offer sustenance while mitigating the symptoms of burnout.

Many on-site foodservice operations have migrated well beyond the simple tasks of providing healthy menu items, nutritional information pertaining to menu options,

BON SECOURS' INTEGRATED WELLNESS PROGRAM (RICHMOND, VIRGINIA)

It's easy to say that an on-site foodservice operation can be more than just a provider of quality food. However, to coordinate with other departments and draw on every potential synergy to deliver additional services—and added value—takes communication, coordination, and, of course, a willingness to try. At Bon Secours Health System in Richmond, Virginia, the foodservice team works with the wellness program facilitators in many ways to deliver a clear message: Wellness matters, and food is a primary component of everyone's health and longevity.

Bon Secours, a four-hospital, 5,000-employee healthcare provider, has always made it a priority to encourage its constituents—customers and employees alike—to take care of themselves. In the mid-1990s, they assumed a more aggressive posture in this area and launched a systemwide wellness program, using a phased approach that began with St. Mary's, the largest hospital in the system.

While many organizations have wellness programs, this one purposefully involved foodservice at a high level. The director of wellness services worked hand-in-hand with the foodservice director to ensure that the two departments worked as one. First, the foodservice management team introduced "heart-healthy" menu items in the main eatery, along with the recipe and nutritional analysis for each item. With the help of the department's dietitians, advice in the form of pamphlets and informational packets was also offered to customers of the main dining facility. These health-friendly entrées were included on patients' menus (where appropriate, given particular dietary restrictions).

Using the foodservice department as a main channel of communication, employees were also invited to join a fitness program involving weekly meetings, a formalized reporting mechanism, and custom-designed exercise and nutritional guidelines. The support from the host organization was so strong that employees were allowed time away from work (but were still paid) to attend meetings. The topics of the weekly meetings ranged from methods of healthy food preparation to the benefits of regular cardiovascular exercise. Graduates of the program were rewarded with a celebratory lunch and a certificate noting their accomplishment.

Granted, such a program is a wonderful benefit. But was the multidepartmental endeavor successful in terms of the bottom line? The first few years of data suggest that the program substantially reduced turnover and decreased healthcare expenses, with the short-term benefits somewhere in the range of a 25 percent reduction in related expenses. The cost of the program, on the other hand, was minimal.

What are the key success factors enabling this program to produce such impressive results? According to everyone involved, the primary factor is commitment. This means considerable time spent together during the planning process. Arguably equal in importance, buy-in from the employees in the related departments equates to a smoother transition. It is vital that everyone understand the benefits of the program so that they can openly support it (and not undermine the good intentions of the leaders). Finally, tracking the results is integral to maintaining upper management's support. After all, while personal wellness is desirable, organizational wellness in terms of financial health is primary, ensuring that everyone has an ongoing position in the organization!

and a reprieve from the chaos of everyday office life. Some have embarked on educational efforts, focusing on modifying less-than-optimal eating habits. Some have aligned with fitness programs, offering meal plans that correspond to preset guidelines, thereby allowing customers to dine in the eatery while still adhering to their modified dietary plans.

Taking it further still one more step, other operators have teamed with wellness programs, most often initiated by the host company's human-resources department. An apt example is the next best practice. The wellness program at Bon Secours Health System in Richmond, Virginia, underscores the important role of foodservice in delivering more than just good food to customers.

CHAPTER SUMMARY

Changes in local, regional, national, and global economies in recent years have led to the elimination of nonessential operating expenses, including those associated with employee perks such as subsidized foodservice. This has placed unprecedented pressure on on-site foodservice managers to reduce expenses while delivering high-quality foodservice. In the healthcare industry, increased operating expenses have forced administrators to look both at new ways to increase revenues and at any possible method of reducing expenses that directly affect foodservice operations.

In addition, host organizations have begun to look at how the foodservice department fits into the overall scheme of operations and how the customer is viewed. In healthcare, for example, the practice of treating patients as customers has emerged. For foodservice operators, these changes have resulted in the need for more creative approaches to managing the business at hand. Yet such changes have not come without challenges; stress and creative tension are pervasive.

Today, on-site foodservice is charged with more than just providing high-quality food at a fair price. It is incumbent on foodservice professionals now to also educate customers, often through menu boards and other media, regarding nutrition and the importance of maintaining good health.

KEY TERMS

daypart	full-time equivalent	creative tension
diagnosis-related groups	patient-focused care	burnout
health-maintenance organization (HMO)		

1. How have changes in the economy affected business? How have they affected on-site foodservice operators?

2. Referring to Motorola's strategy to pit their in-house foodservice division against outside contractors, what does this say about the strategic nature of on-site foodservice?

3. Explain what challenges the healthcare industry has faced, beginning in the 1960s and continuing to the present. What similarities can be found in other industries?

4. Why are capitated reimbursement plans problematic in terms of revenue?

5. If you were asked to reduce your labor force by 10 percent, what positions would you likely cut first? What effects might this have on your operation? What alternatives might you propose if asked to cut labor expenses?

6. How is foodservice management at Ohara different from that in other companies you have seen?

7. Referring to the traditional organizational configuration depicted in Figure 2.3, what do you immediately see as potentially problematic for delivering hospitality-related services to customers?

8. If you had to predict the future, what would the next iteration of patient-focused care entail?

9. Can creative tension be positive? If so, describe a situation in which it might be positively harnessed.

10. How can on-site foodservice providers facilitate the efforts of organizations to minimize burnout and promote wellness?

END NOTES

[1] As an example, in a *Business Week* article in mid-2001, 39 percent of small business owners said they believed the United States was in a recession, while 31 percent said the opposite; 21 percent simply weren't sure. Of those who thought the recession was occurring, more than two-thirds reported that it would likely end within a year. For more on this, see Editorial staff (2001, July 16). How's the economy doing? *Business Week,* 5.

[2] More on creative tension and the change to a learning organization can be found in Senge, P. M. (1990). *The fifth discipline.* New York: Doubleday.

CHAPTER 3

OPERATIONAL CONFIGURATIONS

Changes in terms of host organizations, and the position and role of foodservice within these companies, have contributed to the evolution of on-site foodservice in myriad ways. As discussed in the previous chapter, the dominant changes include the expanded role of foodservice and the way customers are regarded. Many on-site operators have responded to these issues in part by reconfiguring their operations. While the phrase "thinking outside of the box" is overused and certainly passé, it applies here in that managers are rethinking how everything is done, from preparing the food to delivering it to the customer.

Thus, this chapter focuses first on different production systems that are most viable for today's leading operators. In the "old days," say the 1970s, a typical foodservice operation revolved around the kitchen. Through decentralized production approaches, today's eateries may not even have a kitchen. The discussion then turns to delivery systems, a relatively new area of expertise that foodservice managers must master. The chapter concludes by analyzing different service styles. In reciprocal fashion, production, delivery, and service are connected, with each building on the approach taken in the previous stage. This means more options for foodservice managers. It also underscores the fact that on-site operators must exploit opportunities in each of these areas to maximize the functionality and profitability of their units.

PRODUCTION SYSTEMS

Production of even the most basic menu items requires some planning in terms of layout, design, and execution to ensure consistent delivery of the dish to the customer. Take a simple bowl of vegetable soup. The product itself can be made from scratch, using house-made stock, and may include any number of vegetables. It can be purchased from a vendor in a can, in much the same way that people buy canned soup from a grocer and then heat it at home. Or it can be purchased as a refrigerated or frozen product, in which the soup needs heating but no additional ingredients. Similarly, the soup might be made in-house in large quantities, then refrigerated or frozen for later use; it could also be shipped to tertiary sites for use there.

Each of these preparation techniques requires different considerations and offers different options in terms of delivery. Making soup from scratch necessitates stock pots, appropriate ingredients including both vegetables and seasonings, a preparation area replete with prep tables, knives, and so on, the equipment necessary either to put the product into service (in the case of a cafeteria) or to hold it in storage, such as refrigeration space, and, of course, a knowledgeable cook to create the end product. These needs are substantially reduced when using prepared products, whether they are canned or prepackaged in some other way. The question, then, as with most operational decisions, becomes one of cost and quality. Each method offers a different balance of these factors, and each is dependent on the desired outcome.

Decentralized versus Centralized Production Systems

Once upon a time, every foodservice outlet—from commercial restaurant to small hospital coffee shop—was designed with a kitchen in the back of the operation and a dining area in the front. There were good reasons for this: The closer the production area is to the customer, the less degradation of product, including temperature, texture, and other key attributes of food. Furthermore, these conventional operations typically featured **scratch cooking**—using largely unprocessed ingredients in the preparation of menu items.

Thus, these on-site foodservice operations had the same number of kitchens as eateries. This **decentralized** approach facilitated efficient tracking of products and allowed great versatility in terms of menu development. It also meant great redundancy in food-production activities: Lettuce was chopped for salads in kitchen A for outlet A and in kitchen B for outlet B. This became particularly problematic as labor costs escalated over time.

Similarly, because of the increased availability of prepared items, on-site foodservice managers had to think more about possible efficiencies, particularly when multiple outlets were involved. In some cases, much prep work could be eliminated in all outlets with items available from food distributors (e.g., by purchasing boneless, skinless chicken breasts versus deboning and skinning whole birds). In the case of desserts, entire prep areas could be eliminated given the many high-quality prepared desserts available. This trade-off became an elementary decision in most cases, assuming that the quality of the prepared items was similar to that of products prepared from scratch. If the cost of labor was greater than the differential between the prepared item and the unprocessed, raw product(s), then the fiscally prudent choice was to take the road offering greater convenience.

As this mindset become more common, managers also began to look for other efficiencies. Thus was born the **cook-to-inventory kitchen**—an operational arrangement in which **centralized** production services multiple outlets. There are two types of cook-to-inventory kitchens. The first is sometimes termed **ready prepared,** which encompasses cook-chill and cook-freeze methods of food production. The other is **commissary production,** in which food is prepared and shipped. This may include such basic items as shredded lettuce or grated cheese. It may also include prepared dishes, shipped to the tertiary sites in bulk but already heated and ready to serve, or it may be ready prepared and need only heating or final assembly at the destination site. Some commissary production systems involve a hybrid centralized-decentralized approach in which some food is centrally produced and shipped to a secondary location that prepares the remainder of the items or converts the shipped items into finished products.

The cook-to-inventory kitchen offers the key advantages of minimal redundancy and more efficient use of labor. It also integrates new technology with a calculable return on investment, which does not vary according to market forces in the same way as does labor. It affords operators much better control of the entire production process, which translates into improved food quality, more consistency, and enhanced food safety.

This approach to production offers still other product- and labor-related efficiencies. Cost savings are realized through large-volume purchases. Internal controls in a large central production facility are executed much more easily than in several smaller kitchens, thereby reducing shrinkage due to theft. Finally, food can be produced regardless of the daypart for which it is destined. In other words, ribs can be cooked first thing in the morning. Thus, dedicated production personnel can be utilized during a continuous eight-hour period as opposed to employing several highly paid production personnel to cover the three major dayparts.

Each different approach to the cook-to-inventory kitchen offers its own set of advantages and challenges. **Cook-chill** food production is a system by which food is prepared at or near pasteurization temperature, packaged in special airtight bags or casings, and chilled to less than 40°F in less than 60 minutes (sometimes slightly longer for certain items). The prepared items can be refrigerated for up to 45 days in most cases. **Cook-freeze** production, which offers extended shelf life, is similar to cook-chill, except that food products are frozen using a blast freezer or cryogenic freezing system after the initial cooking process.

The method of initial cooking in cook-chill depends on the type of food. Pumpable foods, such as sauces, soups, gravies, baked beans, taco meat, and any other product that will flow when hot, are usually prepared in a kettle and then put into bags using special pumping equipment (for example, see Figure 1.9 in Chapter 1). The bags vary in capacity but typically hold one to two gallons. They are then placed in a tumble chiller, as shown in Figure 3.1, where the bags can be cooled rapidly.

Foods such as roasts, ribs, and various other cuts of meat are often vacuum packed in bags or casings and then loaded into hot-water cook tanks. The hot circulating water cooks the meat slowly, retaining most of the natural juices, and keeps shrinkage to a minimum. After cooking, cold water is circulated in the tank to quickly reduce the internal temperature of the items. The precursor to this practice is known as **sous vide** ("under vacuum"), invented in the 1970s to curb shrinkage of foie gras. Cooks in France found that sealing fresh food in a vacuum pouch, then slowly cooking it at a low temperature in circulating water, offered improved flavor since juices are trapped within the pouch.

Items that are not appropriate for either of the aforementioned methods are sometimes prepared using combination ovens ("combi-ovens") in which convection heating and steam cook the items. After cooking on racks or in shallow pans, the items are transferred to blast chillers, which chill the food quickly using compressed air. The shelf life is shorter than that of items prepared in other ways, such as vacuum-packed soups, but it is still sufficient to serve the purpose.

Rethermalization—the process of reheating the precooked food items to serving temperature—is accomplished in a variety of ways, depending on the product and availability of equipment. Most often, rethermalization is accomplished through the use of kettles, convection steamers, ovens (including conventional or deck, microwave, convection, conduction, cook-by-light, or combi-ovens), skillets, hot-water-bath steam tables, or braising pans. This can be accomplished in large or small quantities, depending on the application.

FIGURE 3.1 Bags of Beef Stew in a Tumble Chiller at MetroHealth Medical System's Central Production Facility, Cleveland, Ohio. Courtesy of MetroHealth Medical System.

CONSOLIDATION AND CENTRALIZED PRODUCTION AT THE MEDICAL CENTER OF CENTRAL MASSACHUSETTS (WORCESTER, MASSACHUSETTS)

Prior to its merger with the University of Massachusetts Memorial Healthcare System, the Medical Center of Central Massachusetts consisted of two hospitals, Memorial and Hahnemann. The Memorial campus featured some 300 patient beds along with tertiary services such as an emergency room, a small assisted-living wing with 30 beds, and modest outpatient services. A little over one mile away, Hahnemann was smaller, with 160 beds, but it also offered various related services. Both facilities had considerable public eating areas, with a large cafeteria and coffee shop at Memorial and

a fair-sized, well-designed public eatery at Hahnemann.

As one might expect, foodservice at each hospital was run autonomously during the first several years in which both hospitals were part of the same system. Over time, the menus were consolidated so that employees and guests of the two cafeterias could enjoy the same menu items. The same recipes were used, and the same price structure was in place. Deliveries of raw products were made to both kitchens. Following several staff reductions aimed at lowering operating expenses, the administration and foodservice management team realized that redundancies in the system justified rethinking the organization and configuration of the systemwide foodservice operation.

A number of factors were considered. First, Memorial had a larger kitchen, with much more storage capacity (for both dry goods and refrigerated items). Second, the equipment in both kitchens was in relatively good shape. Third, a number of employees had been cross-trained to the extent that most members of the production staff were comfortable at either location. Finally, the administration adamantly insisted that it wanted a solution that required minimal capital investment.

Several scenarios were explored. The goal was to maximize efficiency and productivity while minimizing costs, particularly labor-related expenses. At the same time, however, everyone knew that any change would be difficult to implement since most employees had been with the respective hospitals for many years and most had a strong allegiance to their tried-and-true ways of operating. This included nurses as well as employees in other departments directly affected by foodservice. Furthermore, there was a common perception that the food—both patient and cafeteria—at Hahnemann was better.

Thus, the decision makers enlisted the help of many other managers, including representatives from nursing and environmental services. Here's what the group conceived: As for food production, all hot food would be produced in the Memorial kitchen. Cold preparation would be largely performed at Hahnemann, since prep space during peak feeding times was limited systemwide. Delivery would be achieved though a dedicated truck fitted with a custom lift meeting the unique needs of both kitchens' loading docks. As for delivery to patients, the team, through their research, integrated a technologically advanced food cart (see Figure 3.2). Holding 20 trays, each cart was capable of keeping the hot food at a desirable serving temperature through heat conduction on specially designed plates. Cold food was kept cold in a separate compartment and added to the tray upon delivery to patients. Regarding the Hahnemann cafeteria, food was to be trucked over in bulk containers and placed directly on the serving line upon delivery.

The cost of the state-of-the-art carts, related service ware, retrofitting of holding space in terms of electricity needed for the carts, and delivery truck with custom-fitted lift totaled some $100,000. Labor reduction included the foodservice director at Hahnemann (the supervisory staff still supplied coverage throughout all shifts), most of the production staff, and two sanitation workers. These reductions resulted in a decrease in labor expenses, including all labor-related costs, of just over $200,000 in the first year alone. Thus, the payback was accomplished within the first six months of implementation.

Was the new system successful? Absolutely. Patient satisfaction at both hospitals improved dramatically, particularly in terms of food quality and food temperature. Of greatest importance, the perceived quality of the food systemwide was enhanced, with the management team heralded as fiscal miracle workers!

FIGURE 3.2 Burlodge Friotherm Cart. Photo courtesy of Burlodge USA, Inc.

These centralized approaches to food production offer many advantages and underscore the multiplicity of possibilities available for using prepared food items in nontraditional ways. An example of just one of the many ways prepared food items—whether using centralized production or buying premade products—can be used is through a return to the decentralized operational configuration in the form of a **convenience kitchen.** Convenience kitchens allow for the elimination of traditional large-scale production areas (and the requisite traylines in healthcare foodservice) through the integration of small modular kitchens located close to the point of service. In a healthcare setting, this might be on each floor of a hospital. In a B&I setting, it might be part of a small coffee shop.

In most instances, convenience kitchens use microwave ovens to rethermalize food intended to be served hot. Many factors need to be considered, however, before integrating convenience kitchens in an operation. A newly constructed satellite outpatient facility located in Massachusetts, for example, implemented a convenience kitchen, operating under the premise that food could be delivered from the main medical center located 20 minutes away. The tertiary operation was designed with minimal cooking equipment and had no ovens except large commercial microwaves. In fact, the placement of the kitchen precluded any chance of later adding equipment requiring venting (such as a conventional oven). Things didn't go as planned because of challenges associated with food production at the central facility and subsequent delivery. The satellite facility then resorted to using frozen dinners. This worked for some time, largely because the number of customers was small.

The downfall of the facility was the foodservice team's decision to keep the source of food a secret from the rest of the organization. This unfortunate decision led to a system failure on Thanksgiving Day. First, this particular facility was not supposed to have inpatients; patients were expected to go home after their procedures. On this one day, however, a number of patients suffered complications and were forced to stay overnight. The next day—Thanksgiving—the foodservice department admirably wanted to provide traditional turkey dinners to the patients and their families who would likely visit. Unfortunately, the only viable option was Stouffer's turkey dinners.

Initially, the meal period went as planned. The freezer held just enough meals to cover the patients and visitors. Just when the manager thought he had made it out of the woods, the nursing supervisor came to the kitchen to compliment him on the wonderful meal and to ask if the foodservice department might provide a turkey meal for the nursing staff, even if it consisted of leftovers. Serendipitously, the nurse happened to notice the empty frozen food boxes in the trash. Embarrassed that she had unknowingly told patients and their families that the food was freshly prepared, she—according to reports—had the foodservice manager for lunch instead of a turkey dinner.

DELIVERY SYSTEMS

These production systems and approaches allow operators great latitude in deciding how to configure their operations. This extends to how food is delivered, whether to patients, employees, or guests. The advantage is that each approach affords foodservice managers the ability to configure the operation according to the needs of the customer.

Delivery pertains to the transportation of food products to the place of service. As part of the delivery process, food may be prepared and served at the point of service, prepared elsewhere and delivered in a ready-to-serve state, or rethermalized at the point of service. The appropriate type of delivery system depends largely on the production approach, organizational needs, equipment, quality standards, various economic factors, and the acumen of management.

Traditional Delivery Systems

As noted earlier, the **traditional delivery system** in the majority of smaller on-site operations today involves moving food from the kitchen to a serving line. A variation is seen in most restaurants, most notably quick-service restaurants, where the food is batch-prepared, as are hamburgers in a McDonald's restaurant, or cooked to order, as in a Wendy's restaurant. In either case, the food is delivered immediately after production in the kitchen.

This is not to say that food cannot be prepared elsewhere. As in the commissary production approach described earlier, many operations use prepared products such as diced vegetables and shredded cheese. Take a typical Taco Bell, for example. Little prep work is done in the kitchen since most accompanying items are delivered in a ready-for-use state. The production staff is charged primarily with the final assembly of menu items. Thus, tomatoes do not need washing, lettuce does not need shredding, and peppers do not need chopping (since the salsa is shipped ready for use, too).

Modified Delivery Systems

Modified delivery systems integrate preprepared food items, often in concert with cook-chill production or a commissary approach. Food is shipped cold, to be rethermalized at the point of service. Hence, minimal kitchen equipment is needed, except for those appliances needed to reheat the various types of items.

With this approach, quality control is enhanced across multiple units. Returning to the Taco Bell example, this can be seen in the meats used for tacos. The beef and spices are combined and cooked in a centralized kitchen. Containers of the taco filling are then sent, properly chilled, to the units, where the product is reheated to the appropriate temperature. The final step in the delivery system is to assemble the taco.

Envision 100 units working under this production and delivery configuration and the efficiencies are self-evident. The beef is sautéed in huge batches by 1 or 2 employees versus the 100 (or more) who would be required if it were prepared at each site. Moreover, delivery is easily facilitated, since numerous other products must be delivered to each location in a refrigerated state. The final step of rethermalizing is simple and straightforward.

Expedited Delivery Systems

The extension of a centralized approach achieves the goal of minimizing labor in tertiary sites, as can be seen in **expedited delivery systems.** In this system, food is centrally produced, then shipped in portioned, ready-to-eat servings. In the case of correctional foodservice, this would take the form of prisoner trays. In B&I or schools, it would be found in grab-and-go situations. Quality and portion control are maximized, since menu items are fully standardized and the item composition is unaffected by the number of tertiary sites.

The challenge of this delivery approach is to maintain the food's temperature, texture, and moisture control. Once plated, hot food continues to cook. In the process, it can become dry, overcooked, or, in extreme circumstances, virtually inedible. Hence, the problems in the delivery process, which are uniquely difficult for this approach, are to balance maintaining proper temperature with managing humidity control in whatever equipment is used during transport. Some foods, such as those covered in sauces or gravies, are less problematic; others, such as fried food or naturally delicate cuts of fish, deserve extra consideration. Following is a best practice that demonstrates the plausibility and advantages of just such an approach.

BEST PRACTICES IN ACTION

ADVANCED MEAL DELIVERY SYSTEMS AT METROHEALTH MEDICAL CENTER (CLEVELAND, OHIO)

Located on Cleveland's near-west side, an economically depressed area, MetroHealth Medical Center is quite accustomed to seeking unique solutions to financial challenges. Like most healthcare systems in the 1990s, this public regional provider faced declining revenues and increased costs. However, while similarly positioned institutions engaged in stringent cost cutting, particularly in such support areas as foodservice, MetroHealth applied the lessons learned from watching the downfall of other organizations. Taking the more difficult road, MetroHealth began looking at systemwide changes that, while requiring considerable investment in the short term, might produce profitable results over time. This was particularly difficult given the expansion of the organization to its current configuration of three centers totaling 750 patient beds.

In its original form, foodservice at MetroHealth featured a traditional cook-to-serve trayline system, which was staffed with a 12-hour-per-day unionized labor force. The main kitchen encompassed a generous 15,000 square feet and was located near a large-capacity dining area. Prior to the many impending changes, the foodservice departments of the three core facilities had some 275 FTEs on the payroll.

Seeking efficiency and economy, the foodservice management team began the task of investigating options. The discussions were long and sometimes heated; they included MetroHealth's administration and key members of various departments. The various constituents also used their industry contacts to explore the many options available. In terms of feasibility and return on investment, they identified the cook-chill system as the most viable solution.

In tandem with this decision, the team developed a detailed two-phase integration plan. The first phase involved the installation of the majority of equipment needed for cook-chill production, including an expanded refrigerated storage area. This would allow the department to produce some 60 percent of the system's patient, employee, and guest menu items. Phase two—the more capital-intensive phase—required converting the old production area to a temperature-controlled prep room with a centralized cold trayline and a production area capable of supporting the entire MetroHealth system; it would also be sizable enough to produce excess capacity—items that might be sold to other institutions at a profit.

The final renovations, which totaled nearly $3.85 million, were completed just before the birth of the new century. The payback is impressive: In comparing the department's financial operating figures with those of the newly configured operation, the savings realized an average of $1.15 million per year, resulting in a payback period of just under four years. Moreover, these savings will likely be even greater in future years, given the tight labor market and increasing labor-related expenses.

How are these savings possible? The efficiencies stem from labor reductions and better utilization of food products achieved through the new production system. The majority of food is typically prepared 18–20 days in advance by two FTEs. As shown in Figure 3.3, food is held in large walk-ins until needed. Trays are assembled, with the food still cold, in the temperature-controlled area, as depicted in Figure 3.4, and then loaded into dual-zone rethermalization cabinets where items to be served hot, including beverages, are brought to serving temperature while cold items are kept cold.

But what about the patients' perceptions of the food? Quarterly data for the several quarters since the renovation was completed reveal that patient satisfaction is substantially higher than it was before the system was established. Moreover, it appears that things continue to improve, judging from the multidimensional satisfaction scores, as the food-service team becomes more adept at utilizing the various aspects of the system.

Cook-chill production is also used for food in

FIGURE 3.3 Food Items Prepared Through the Cook-Chill System Are Stored in the Large Walk-in at MetroHealth Medical System's Central Production Facility, Cleveland, Ohio. Courtesy of MetroHealth Medical System.

FIGURE 3.4 Tray Assembly, Using Cold Food That Will Be Rethermalized Just Prior to Service, at MetroHealth Medical System's Central Production Facility, Cleveland, Ohio. Courtesy of MetroHealth Medical System.

the various public outlets. Owing to the lack of shrinkage, along with consistency in quality achieved through the system, food cost is reduced while quality is improved. And because of the economies of scale realized through producing similar items for multiple outlets, the purchasing and preparation processes have been streamlined. What's also impressive is that the foodservice management team projects that the current production capacity could be almost doubled—with no additional equipment—as MetroHealth continues to expand.

SERVICE STYLES

As with the many types of delivery approaches, service styles vary, depending on the method of production, and also are subject to the delivery system in place. Granted, the common goal is to ensure maximum customer satisfaction, but the challenges inherent in meeting this goal can be numerous and daunting. The biggest challenge is to make sure that the menu items created through the production process and delivered to the point of purchase retain their quality.

Some have examined the selection of the service approach on the basis of utility and pleasure. Take vending machines, for example. The utility is very high: The customer puts the money in, makes the selection, and receives the food product, a very efficient process. Aside from the visceral satisfaction produced when eating, however, the personal element is nonexistent. On the other end of the continuum is tableside service, in which the server finishes items while providing an interpersonal element to the exchange. The customer and server are fully engaged; the customer might ask about preparation or might suggest ways in which the server can make the item exactly as the customer wants it. In turn, the server focuses on the one customer. Is this efficient? Not in the sense that it optimizes the exchange. On the other hand, however, this type of service is arguably most pleasurable and produces a very positive outcome. Thus, the type of service style is a derivative of the restaurateur's goal, which is in turn a response to customers' needs, wants, and expectations.[1]

Staying with the notion of balancing utility and pleasure, then, service styles can be narrowed to five primary categories:

1. Self-service
2. Portable meals
3. Tray service
4. Wait service
5. Tableside preparation

Self-Service

Self-service is the simplest approach. As noted earlier, vending is a good example of self-service simplicity. Other types of self-service include cafeterias, buffets, and salad bars.

The beauty of vending is that service is available 24 hours a day. Decades ago, vending was limited to candy bars, sodas, and other convenience items. (Think back to the penny gumball machines from the early twentieth century.) The next generation of vending machines featured frozen confectionary items and prepackaged sandwiches. Today, machines are available that offer freshly baked chicken potpies, French fries, and even spaghetti with meat sauce, which is cooked to order! In Japan, machines that offer dried squid, sake, and small kegs of beer are commonplace. In Belgium, automated convenience stores are encased in single machines that contain some 200 different items, including many food items.

There is no arguing about the convenience these machines afford customers. The challenge for operators is the cost and upkeep. The good news is that some machines can store products for up to four months, even when the item is automatically cooked to order. The downside is that some machines with state-of-the-art technology can run more than $50,000; repair of these machines can be equally expensive owing to their level of sophistication.

Cafeterias represent the next step in self-service. The traditional straight-line-counter cafeteria remained unchanged for many years. In universities and office complexes, the line was set up with desserts first, followed by salads, side dishes, entrées, and beverages. The only variation was a choice between "all you care to eat," in which the customer paid at the beginning of the line, and "pay per item," with the register at the end of the line. The layout was so standardized that an early textbook even featured this format as the "conventional arrangement."[2]

There is no longer a standard approach to the cafeteria-style layout. The straight-line format, considered obsolete because of its inefficiency and intrinsic inability to merchandise with maximum effectiveness, is often replaced by the scatter system. This can be a simple variety of stations located strategically around the area or a composite of different miniature outlets that resemble a small food court.

Salad bars and smorgasbord-style offerings are the final type of self-service and often complement other configurations, including cafeterias and wait-service-type approaches. In many commercial restaurants, for example, a salad bar may be an included feature of all entrées. Many on-site operations expand the concept with such items as top-your-own baked potatoes, taco bars, ice cream sundae bars, and popcorn bars (where customers add their own choice of flavorings to prefilled bags of popcorn).

Portable Meals

Portable meals are similar to self-service in that the menu items are usually ready to eat or need only reheating. This can include **home-meal replacements** (HMR), prepared items that customers purchase so as not to have to cook for the family when they arrive home. Quick-service chicken restaurants such as KFC are an example of producers of HMR even though they embraced the concept long before the term was coined. Such chains market chicken on the bone (fried, baked, etc.) accompanied by limited side items. In a similar fashion, on-site foodservice operators often tap into HMR as a venue for increased service and profits. Such portable meals can be as simple as boxed pizza or as complex as complete, multicourse holiday dinners.

Another category of portable meals comprises delivered items. In B&I, for example, many customers want food delivered to their offices, a good option for operators since it means that the customer is spending the money on-site versus calling outside for delivery. In this way, portable meals as a service style mean capturing more **stomach share,** an attractive business outcome.

These delivered meals can also be found off-site, where the on-site operator makes deliveries beyond the immediate marketplace. A good example is a university foodservice operator who delivers food to professors' offices or even to students' apartments off campus. The key benefit of this style of service is the expanded customer base.

Tray Service

Tray service involves meals or snacks that are amassed and carried to customers who, for whatever reason, cannot or choose not to eat in the common dining area. This service style is often seen in hotels; better known as "room service" in such settings, it is simply food served to the customer in a guest room, meeting room, or poolside on a tray or cart. Tray service is most common in healthcare foodservice, where it is like hotel room service without the service charge.

The greatest challenge with tray service is maintaining quality. As discussed in the section on delivery, there are many challenges, not the least of which is maintaining temperature, texture, and moisture content. Each of these attributes is typically affected negatively by time. Thus, it is incumbent on the operator to minimize the time the food is on the tray before reaching the end user.

Wait Service

Wait service, which can include table or counter service, can be very simple or fairly elaborate. With counter service, the employee serves as an intermediary between the kitchen and the customer. This can be as simple as serving food from a steam table. The difference is the personal touch. Referring back to the notions of utility and pleasure, this style of service introduces more pleasure through personal interaction than is found in self-service, even if the food is similar.

In more elaborate situations, such as when the foodserver both takes the order and delivers the food to the table, the pleasure side of the equation is even greater. In such service approaches, more formality is involved. The key is that there is an intermediary who, by design, is able to increase customer satisfaction.

More and more on-site operators are realizing the profit potential of integrating outlets featuring varying levels of wait service. Granted, the application is site-specific and the challenges are inherently greater. The potential, however, is considerable given the operators' prowess in traditional on-site styles of service. This is just another way in which service can be a differentiator; it can also send a message to customers that they have more options and do not have to look outside for quality food and service.

Tableside Preparation

Tableside preparation features food prepared or at least finished at the table in front of guests. The most common example of this is the preparation of a traditional Caesar salad. The foodserver prepares the dressing in a wooden bowl, often adjusting the recipe to the voiced requests of the table's occupants. Other ingredients are added, and the salad is tossed. The foodserver then apportions the salad accordingly and serves the guests.

This style of service is seen more and more in such on-site applications as sports venues that have private rooms or "boxes." It is also found in upscale executive dining rooms or at special catered events in any of the on-site segments. Granted, its application is somewhat limited due to the labor issues (such as cost, training, etc.) and the necessity of some specialized equipment, including but not limited to at least a guéridon (French for "pedestal table"—the term is used to describe a rolling cart used for preparing foods tableside) and a réchaud (French for "reheat"—the term is used to describe a chafing dish). Nonetheless, it is a viable service style for on-site operators given the right combination of circumstances.

There are myriad service styles, and managers often adopt more than one for different facets of the operation. The key is to tailor the style of service to maximize the potential of the outlet. In turn, delivery and production must coalesce and complement the service approach in such a way that the creation and provision of foodservice are seamless. The real test, after all, is whether customers' expectations can be continually exceeded. If the answer is yes, then the recipe for production, delivery, and service approaches is correct.

CHAPTER SUMMARY

Production systems are predicated on several factors, including considerations regarding equipment, technology, and labor. Decentralized production, in which the servery is attached to a dedicated kitchen, is historically the most common arrangement in foodservice. Some decentralized operations use prepared food items to decrease their reliance on labor, thereby trading higher food costs for lower labor expenses.

Centralized production, usually accomplished through cook-chill methods, achieves the objective behind cook-to-inventory kitchens. This approach allows a single production facility to service multiple outlets; substantial economies of scale are achieved since there is minimal redundancy in the production process. Food produced in a centralized kitchen is commonly rethermalized at the point of service. Like the production method itself, the methods of rethermalization depend on the products and availability of equipment.

Delivery systems, which involve the transportation of food products to the point of service, vary according to operational configurations, the production system, and the intended style of service. With traditional delivery systems, food is simply moved by an employee from the kitchen to the front of the house. With modified delivery systems, food is shipped cold—sometimes at considerable distances—and then is rethermalized. Finally, expedited delivery systems involve shipping food in a ready-to-eat state.

Service style, categorized into the five primary classifications of self-service, portable meals, tray service, wait service, and tableside preparation, is the final piece of the puzzle that crowns the production and delivery processes. As such, the style of service must complement the other aspects of the operation. The challenge in merging the different components of the operational configuration is maintaining food quality.

KEY TERMS

scratch cooking
decentralized food
   production
cook-to-inventory kitchen
centralized food
   production
ready prepared
commissary production

cook-chill
cook-freeze
sous vide
rethermalization
convenience kitchen
traditional delivery
   systems
modified delivery systems

expedited delivery systems
self-service
portable meals
home-meal replacements
stomach share
tray service
wait service
tableside preparation

1. Given the different production systems available, why would you select one over another?
2. The cook-to-inventory kitchen has been hailed as the best approach for on-site foodservice. Do you agree? Why or why not?
3. Upon its adaptation in different countries, one of the fears of using sous vide as a production method was that it would decrease food safety. Is this a valid concern?
4. What is a convenience kitchen? What is a good application for it?
5. Outline the necessary steps for moving from a traditional delivery system to a modified delivery system.
6. How might the primary problems associated with expedited delivery systems be solved?
7. The advanced meal delivery system at MetroHealth Medical Center is the picture of efficiency. Why would someone not want to take the same approach?
8. Plot each of the service styles on a two-dimensional grid showing the level of utility and pleasure for each.

END NOTES

[1] For more on this topic, see Walker, J., & Lundberg, D. E. (2001). *The restaurant: From concept to operation (3rd ed.).* New York: John Wiley & Sons.

[2] For a good portrayal of the roots of cafeteria service and other related information, see Stokes, J. W. (1982). *How to manage a restaurant or institutional foodservice.* Dubuque, IA: William C. Brown Company.

PART II

FOCUS ON EXTERNAL CUSTOMERS

Building on the foundation developed in the first three chapters, this part takes a more focused approach to on-site foodservice operations, spotlighting issues related to the primary end user. First, Chapter 4 addresses the cafeteria and its customers, focusing on typical challenges such as layout and design, marketing and merchandising, and the effective use of branding. The discussion appropriately looks at these issues not from a functional standpoint only, but also from the perspective of the customer.

Chapter 5 talks about the patient as customer from a critical perspective, building on the discussion of the importance of focusing discretely on the end user. The focus on patients as customers raises some issues that are unique to healthcare foodservice, but the discussion nevertheless has direct implications for other on-site segments. The discussion also illustrates a changing situation: No longer are patients expected to predict what they will want to eat days in advance. Some room-service programs, in fact, are so similar to traditional hotel room service that the guest cannot tell the difference. Best practices underscore innovations in this area, including one that offers food on demand for patients regardless of the time of day. Another illustrates the advantage of personal service in the order-taking process.

Chapter 6 serves as an appropriate capstone to the discussion of external customers. In much the same way that an elegant dessert completes a formal dining experience, catering and special functions serve as the pièce de résistance for on-site operators who want to showcase the talents of their employees. Catering does not, however, create goodwill only within a host organization; it can be used to further the reputation of the foodservice department throughout the community. Arguably of greatest importance, though, it is a wonderful way to boost the bottom line, particularly when labor costs are minimized through cross-training of existing staff members.

In sum, this part continues earlier discussions regarding the importance of viewing every operational issue from the standpoint of the interests of the customer. While the adage "The customer is always right" may not always be true, the concept that the customer should be the focal point of foodservice operations is right on the mark.

CHAPTER 4

STAFF DINING AND PUBLIC FEEDING

The main eatery is a common thread running through all on-site foodservice outlets. Even in corrections facilities, there is a main dining hall in which "customers" congregate to receive and eat their meals. In entertainment venues such as stadiums, the main eatery takes a different form, since most customers sit in the stands to eat. Nonetheless, the notion of food outlets combined with large seating space is one that deserves considerable attention because of its importance to the overall viability of the business.

This chapter begins by exploring how layout and design impact the flow of both products and customers. Many unique factors must be considered when redesigning an eatery; the discussion attempts to highlight those that are most important from an operator's perspective. Marketing and merchandising concepts that complement the design follow. Some concepts are easily marketed but lend themselves less well to point-of-service merchandising. This is a key concern and deserves its due. A corresponding best practice underscores the financial impact of a well-researched design and marketing package.

No discussion of how customers react to design, marketing, and merchandising would be complete without a thorough exploration of branding. The results of effective branding have become more commonly known in the last several years. Still, only the best operators understand the potential of blending different types of brands and concepts. One such operation is highlighted and serves as a best practice owing to its refined approach in this area.

The chapter is intended to speak to on-site operators in every segment. Solutions associated with intelligent layout, integrated marketing, thoughtful merchandising, and artful branding are universal. Of course, it takes the right blend of management skill to excel in all these areas!

DESIGN AND LAYOUT

The process of designing the functional areas of an on-site foodservice outlet is complex. Entire textbooks are dedicated to this topic.[1] Here we focus on three areas that are critical whether one is starting from scratch or thinking about renovating an existing area: planning, operational factors, and layout characteristics. These topics are also useful in understanding the complexity of the overall design and layout process.

Planning

Every on-site operation has a different set of operating goals that affect everything from planning to execution. For example, suppose that an eatery in a business park is intended to provide foodservice only to employees of the host company. Planning functions focus on employees in terms of convenience, service level, production complexity, and flow, while considerations for visitors become minor. This case is very different from that of an eatery in a hospital that is, say, expected to collect 60 percent of its revenues from hospital employees and the remainder from customers such as those visiting patients.

These operating goals frame decisions regarding production and service. In particular, they aid in defining the primary and secondary markets, menu mix, employee skill levels, and levels of line and executive management. The goals also aid in approximating space considerations.

The plan, whether for new construction or renovation, usually evolves in two stages. The first is conceptual; management thinks about what is necessary, what is highly desirable, and what would be merely nice. The second involves the development of a physical plan, replete with drawings, standards, and specifications. Both processes take time. The clarity of vision informing the first typically determines the accuracy achieved in the second.

For large-scale projects, most agree that a **feasibility study** is worth the time and expense. A feasibility study is an analysis, most often conducted by a third party, that indicates whether a project is reasonable, whether it meets the needs of the **target market** (the group of consumers to whom the foodservice operator wants to appeal), and whether it is financially viable, among other things. The study is most often conducted only after the first stage of the planning process is adequately developed.

For smaller projects, the planning process is similar to a capital budgeting process. That is, planning decisions are based on prioritized criteria. Is the need underlying the project urgent, essential, economically desirable, or generally desirable from a return-on-investment perspective? Data for such decisions are both objective and subjective; however, objective evaluative procedures usually are more persuasive when presented to those who control the purse strings.

A common approach to objectively quantifying related decisions is the calculation of the rate of return generated by the investment or **internal rate of return** (IRR). In using the IRR model, the net present value of cash flows is set at zero using an appropriate discount rate. The formula is

$$0 = \frac{CF_1}{(1+r)} + \frac{CF_2}{(1+r)^2} + \frac{CF_n}{(1+r)^n} - PC$$

where

$$CF = \text{cash flow}$$
$$r = \text{actual internal rate of return}$$
$$PC = \text{project cost}$$

A project is accepted if the actual IRR is equal to or greater than the desired minimum IRR, which is also commonly called a "hurdle rate."

For example, suppose that an operator wants to redesign the dining area of a cafeteria. This would require the purchase of new furniture, such as tables and chairs. It might also involve changing lighting, replacing flooring, painting walls, adding new artwork and signage, and building in new waste receptacles. Assume that this project is projected to cost $500,000. In addition, assume that it is expected to increase revenue by $135,000 for the next five years (the depreciation period for the furniture, fixtures, and equipment). Finally, assume that the host organization expects a minimum return, which they target as their hurdle rate, of 15 percent. Using the aforementioned formula, the IRR of the project is 11 percent. (That is, the actual rate of return that produces a net present value of cash flows is zero.) Hence, the fiscally prudent course would be to rethink the project.

It should be noted that IRR is a time-consuming calculation when performed manually, since calculations are made by trial and error. (Various discount rates, r, are tried until the approximate net present value is found to be zero.) Most spreadsheet software and some multifunction calculators include an IRR function, making the process much easier.

Operational Factors

The operating goals determined during the planning process relate to a variety of operational factors. As mentioned, the first concerns defining the primary and tertiary markets. Operations that inaccurately determine their target market experience a host of operational problems down the road and usually have great difficulty adjusting the associated operational characteristics.

With the market sufficiently defined, the **menu mix** is the next factor deserving consideration. The menu mix is a function of projected consumer demand or the popularity of menu items; a proper mix should meet the demands and expectations of the greater part of the primary market. A related issue is pricing. Again, this involves the competitive positioning of the operation in terms of market mix as well as product cost, labor issues, and production methods.

Owing to different subsidizing strategies adopted by host organizations in the various on-site segments, the art and science of pricing is particularly difficult to discuss in specific terms. Most conventional pricing approaches used in the commercial restaurant industry are not universally applicable. That being said, the discussion of pricing in Chapter 6 may help those who want to integrate quantitatively based pricing strategies as part of the menu mix development process.

The skill level of employees correlates directly with the production method, which is a function of the menu. This is an important consideration, and it underscores the importance of appropriately identifying the target market and matching the menu mix to that market; production decisions, including those that account for the skill level of employees, flow *from* these decisions, not the other way around.

Finally, the configuration of line and executive management must be considered as the last operational factor. For example, is a trained chef required? How much supervision is needed in the front of the house? Will management need to use computer-driven heuristics to analyze sales data on a regular basis? This might require a different type of managerial acumen from that which may be sought for a new operation. Finally, and this pertains more to the discussion elaborated in Chapter 10, what management style will be most appropriate for the given type of operation?

A host of secondary operational factors, which form corresponding subsets of the aforementioned items, are also pertinent. These might include food safety and sanitation, worker safety, customer safety, lighting, acoustics, and inventory management. While each of these is important, all related issues must be considered but must be predicated on the primary factors.

Layout Characteristics

Most cafeteria-style eateries are intended to service large numbers of customers quickly and efficiently, with advanced bulk-production techniques and limited staff, which in concert equate to lower variable expenses. The layout must be flexible in that it should accommodate foods served during different dayparts, and must not be too delimiting since it must also be adjusted to serve fewer customers during between-meal periods with, it is hoped, fewer staff members.

While back-of-the-house layout is very site specific—since it is a function of menu-item production needs, type of equipment, and production approaches (see Chapter 3)—front-of-the-house layout is a function of projected sales volume, menu complexity, and desired traffic flow. Each of these is, of course, subject to space considerations and availability.

For operations with low volume and limited menu complexity, a single serving line akin to what once characterized old-style cafeterias may be sufficient. A modern interpretation of this layout type is commonly used in modern coffee houses. Such a design minimizes the need for numerous front-of-the-house employees, integrates a showcase approach to presenting menu options, and is reasonably efficient in terms of traffic-flow management. In most cases, a benchmark for speed of service is anywhere from three to eight customers per minute, depending on the number of selections, promptness of service, and efficiency of each facet of the line (consider the effect on overall efficiency of a slow or error-prone cashier).

Where high volume and increased menu complexity are proposed, a **scatter system** is gaining popularity. An iteration of the food court concept, a scatter system features a marketplace with different stations "scattered" throughout the area. (See Figure 4.1.) Most scatter-system applications integrate varying degrees of **exhibition cooking,** wherein cooks prepare items to order in front of customers. In some instances, point-of-sale terminals are integrated at each station, allowing customers to purchase items from any single station. Because this arrangement is not very customer-friendly if someone wants a slice of pizza from the pizza station and an ice cream sundae from the dessert area, some systems allow single checkout, with cashiers stationed near the exit of the food court area. For "all-you-care-to-eat" concepts, cashiers are located at the entrance.

Scatter systems appeal to most customers, are generally associated with higher perceived food quality, facilitate heavy traffic flows, maximize menu offerings, support faster rates of service, and are flexible. The downside is that they can be extremely labor-intensive. Also, conversion from a traditional single serving line to a scatter system requires considerable capital expense, making the planning phase, consideration

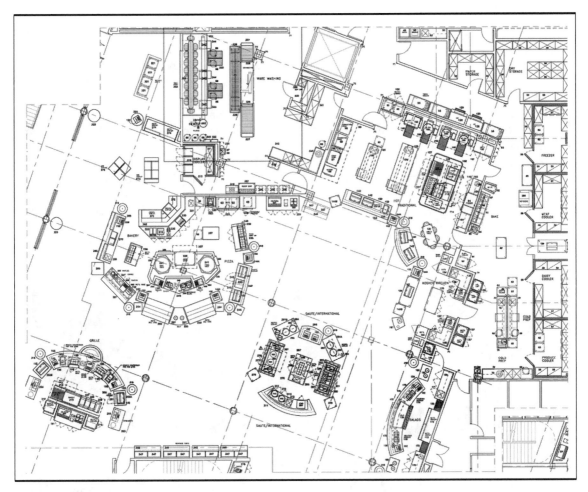

FIGURE 4.1 Draft Schematic for Cornell University's Northstar Dining Hall. Copyright by Dagit Saylor Architects.

BEST PRACTICES IN ACTION

DINING IN STYLE AT CHILDREN'S HOSPITAL BOSTON (BOSTON, MASSACHUSETTS)

For 12 years running, a *U.S. News & World Report* survey has rated Children's Hospital Boston as the best hospital specializing in pediatric care in the nation. The 325-bed comprehensive pediatric healthcare center is the largest in the United States, recording some 18,000 inpatient admissions each year and offering more than 150 outpatient programs and emergency services for more than 300,000 patients annually. Children's is the primary pediatric teaching hospital of Harvard Medical

School; its clinical staff includes approximately 800 active medical and dental staff members as well as 700 residents and fellows. In total, the institution involves about 3,300 FTEs and 850 volunteers.

Regardless of the program or service, Children's most noteworthy attribute is an overarching focus on quality and customer service and a pioneering approach to pediatric healthcare. Children's physicians were the first to use intraoperative magnetic resonance imaging during surgery as a guide for removing a brain tumor in a pediatric patient. The hospital also was the setting for a rare nerve transplant in a 1-year-old girl, the youngest nerve recipient ever; it was the first transplant of nerve fibers from a parent to a child.

Nowhere is the emphasis on customer service more readily apparent than in the café at Children's Hospital. The marketplace concept supports a diverse menu targeted to the local market. This market includes employees, visitors, and customers from nearby businesses. In fact, on any given weekday, the typical customer mix is remarkably similar to that of any other downtown restaurant.

The recent renovation that transformed the eatery and adjoining kitchen totaled approximately $9.6 million, as shown in Figures 4.2 and 4.3. What does that mean in terms of sales? Initially, the renovation equated to a 50 percent increase in revenue. This has leveled off to an impressive 40 percent in-

FIGURE 4.2 The Entrance to the Café at Children's Hospital Boston. Copyright by Sodexho USA Children's Hospital Boston, Massachusetts.

FIGURE 4.3 One of the Many Popular Stations in the Café at Children's Hospital Boston. Copyright by Sodexho USA Children's Hospital Boston, Massachusetts.

crease over the previous year's numbers, with sales during the week of some $12,000 per day.

Unlike most projects, however, this one was not undertaken solely for the purpose of enhancing top- or bottom-line figures. The primary impetus for the renovation was to facilitate the delivery of world-class foodservice to parallel the host organization's reputation for world-class pediatric healthcare. In fact, the feasibility study was conducted not so much to gauge the financial viability of the project but to identify the target market beyond employees and patients' visitors. Tangible return on investment of the foodservice department as an independent business unit was not even calculated.

While the layout and design of the café are impressive, key features embedded in the operational-

ization of the concept make this a best practice. From a production standpoint, several of the exhibition-style stations within the café double as production stations for the patient-service production area. For example, when an order is received for chicken nuggets for a patient, the order is automatically relayed to the fry station in the café. The cook prepares the item while also preparing items for customers in the café. The employee then plates the order and uses the built-in pass-through window to the back of the house, where an expeditor assembles the complete patient tray. During peak dining times, more labor is added to the dual-purpose stations; however, when the peak has passed, the labor is redistributed for maximum efficiency, eliminating duplication of production stations and reducing labor costs.

Perhaps more impressive is the integration of equipment that makes possible the preparation of menu items featured today as well as those potentially integrated into the menu mix of the future. For example, the pizza station features a Roto-Flex oven, which is the highest-producing pizza oven available, as measured by square feet of floor space. The oven decks rotate horizontally, aiding the production process. In addition to baking pizza, the single piece of equipment can be used to bake breads and calzones and to finish a wide variety of other products.

The tapenyaki-style grill is designed with the flexibility to prepare menu items—to order—for all dayparts. For example, the grill can be used to make scrambled eggs and pancakes for breakfast, stir-fry for lunch, and Parisienne sole for dinner. The grill also enables the talented cooks to prepare some impressively true tapenyaki displays of culinary prowess. The countertop induction burners at various other stations are integrated so that they can be moved or replaced with other equipment should changes in the menu dictate. In every facet of the operation, the selection and integration of equipment were made to accommodate changes well into the future, making the operator's job much easier and the service to the customer much greater.

of operational factors, and evaluation of potential layout characteristics all the more important.

An example of an operator who understands the importance and intricacies of eatery design and layout is featured in the associated best practice, "Dining in Style at Children's Hospital Boston."

MARKETING AND MERCHANDISING

At the heart of every successful restaurateur is the ability to market and merchandise effectively. Beginning with self-promotion, most foodservice managers understand that marketing is the key to landing a good job or securing capital from investors. What many forget, however, is that while design and layout are critical, effective marketing and merchandising often dictate whether success will endure.

Marketing

All organizations, regardless of the industry, engage in two activities: They produce a good or service and they market it. Failure to do either one well typically results in failure. This is particularly true in the on-site foodservice industry. After all, many operators can prepare good food. Some can even do so cost-effectively, finding that customers are willing to pay a price that produces an adequate profit. Some have mastered the service of these well-prepared, profitable items. However, it is only those who can then market these items and the operations in which they are served who succeed.

According to the American Marketing Association, **marketing,** broadly defined, is the process of planning and executing the conception, pricing, promotion, and distribution of ideas, goods, and services to create exchanges that satisfy individual objectives. For foodservice managers, this process is accomplished through the development and proper execution of a **marketing strategy,** which involves two primary steps. The first is the identification of the target market, which is the fundamental precursor to most foodservice-business components (as illustrated in the early part of this chapter). The second is to design a **marketing mix** to reach that target.

The marketing mix is the blending of the key elements that satisfy the identified market segments to such an extent that customers will make the desired purchase decision. This jargon from the marketing literature equates to an objective of being foremost in the consumer's mind when the purchasing decision is first considered.

These key elements can vary widely but are commonly divided into four areas: product, pricing, distribution, and promotion. The product pertains to both food and service. What type and style of food does the target market enjoy? This is the same consideration discussed under "Operational Factors." Pricing is one of the most important decisions for on-site operators, given the competitiveness of the marketplace. The key here is using a pricing strategy that succeeds in communicating to the customer that value is high. This does not necessarily mean that an operation must always be the low-cost provider. If someone sells a turkey sandwich for $5 when the deli down the street sells it for $3.50, then it is up to the operator to embed the message through the marketing strategy that the value of the more expensive sandwich is disproportionately greater.

Distribution, discussed in Chapter 3, involves marketing in a variety of ways. If customers believe that menu items are prepared on-site, they may perceive that the quality is better. Thus, it may be important to deemphasize methods of distribution. For other items, it might be advantageous to emphasize a single aspect, or at least a key ingredient, as an appealing differentiator. (This relates to branding, discussed later in the chapter.)

The promotional strategy involves personal selling, sales promotion, and other related communication tools. The key to this strategy is effective merchandising, which contains its own set of challenges.

Merchandising

While some might say that marketing and **merchandising** are separate concepts, for on-site operators trying to maximize point-of-service marketing, these terms can at times be synonymous. Merchandising, by definition, is the act of portraying products and services at or near the point of service that promote the overall business, whether in part or as a whole. Merchandising can be quite comprehensive, embracing market research, product development, advertising, and selling. The problem is that too many operators think that marketing consists only of sending printed materials to customers or using location-specific Web pages creatively and forget that merchandising is just as vital.

The most basic example of merchandising is the garnishing of food, whether on individual plates or platters of food. The goal of garnishing is, after all, to promote the item through enhanced eye appeal. Taken to the next level, garnishing of entire tray-lines is designed to increase the overall appeal of the diverse offerings.

Merchandising is far more encompassing than just garnishing, however. It extends to the cleanliness of the eatery, the friendliness of the employees, and the competence of the entire staff (including management). While these issues manifest in different ways, they are all point-of-service messages that communicate something to the customer, regardless of whether this message is intended.

Like most issues pertaining to a successful on-site operation, menu planning is at the core of a successful merchandising campaign. Assuming that the menu mix is appropriate for the target market, merchandising-related decisions that pertain to all menu items such as color combinations, shapes, and forms of items intended to be served together must be made. In addition, the manner in which food is to be arranged on the plate, serving case, or buffet line must be considered.

Merchandising efforts are subverted by poor sanitation, inferior food quality, inappropriate layout, or inadequate service. Plates on which food has been sloppily apportioned or signage with misspelled words are wonderful examples of negative merchandising. Another example, which too many of us have experienced, is the entrée that has been poorly produced but is garnished elaborately. When more time, money, and effort are spent on the merchandising—in this case the garnish—than on the item that is supposed to be profiled, customers experience dissonance, if not complete dissatisfaction. Thus, while merchandising is an amazingly effective tool for increasing appeal, it can also be detrimental if poorly executed.

BRANDING

In foodservice, **branding** is the inclusion of brand-name products or concepts in the menu or product mix. A branded product or concept is one that is easily recognized by customers who associate it with key features, flavors, or other attributes. Branding is about image; the goal of any brand is to communicate an identity that customers will embrace.

Brand names originated with merchandise but quickly spread to the service industry. Once a tool that only marketers truly understood, branding of any sort is so pervasive in today's society that the lack of a name is at times more conspicuous than its presence. To maximize its utility, however, foodservice operators must understand the subtle as well as the overt aspects to implement branding effectively.

Harvard University Dining Services provides a rich example of how brand identity can affect on-site foodservice. For some time, the catering efforts of the university were less than stellar. Customers perceived the offerings as repackaged dining hall food. The food was good, but not something local residents wanted at special functions. As a result, the catering department commonly lost business to local caterers for jobs both on and off campus.

The on-site management team decided to rebuild the brand, appropriately called Crimson Catering, and began marketing the concept with a new image replete with a new logo. In essence, they were giving their catering arm an identity, one that stood apart from the core business. It wasn't all smoke and mirrors, however. They hired an executive chef, focused on training the line staff, developed specialized catering menus, redesigned all related materials, and marketed the business heavily throughout the community.

For the first time, customers began to associate Crimson Catering with upscale dining; they linked it not to the dining halls but to the reputation of the world-class educational institution. During the year following the introduction of the new concept, sales increased by more than 37 percent, with annual revenues approaching $2 million. Of greater importance for long-term revenue generation, Crimson Catering has become the preferred caterer for large, upscale functions on campus as well as in the surrounding communities.

Branding Basics

In terms of application to foodservice management, there are four types of brands. First are **national concepts.** These have nationwide, and sometimes multinational, name recognition. Leaders in the quick-service restaurant segment are good examples: McDonald's, Burger King, and Pizza Hut, just to name a few. **Regional concepts** enjoy substantial brand awareness but only at the regional level. Residents of Southern California are very familiar with In-N-Out Burgers but would likely be unfamiliar with Legal Seafood, located throughout metropolitan Boston, or Pollo Campero, a chicken brand popular in Costa Rica.

Signature brands, or in-house brands, are those developed internally. The larger managed-services companies have a variety of signature brands, some of which generate considerable brand recognition. For example, in the early 1990s, ARAMARK often integrated their signature pizza concept, Itza Pizza, into most of their accounts. This signature brand became the third largest pizza concept in North America, as measured by annual sales.

The fourth type is **manufacturers' brands.** Manufacturers' brands are most commonly used to accent a relatively standard menu item by giving it added market appeal. Examples are sandwiches with Grey Poupon mustard, burritos with Tabasco pepper sauce, ribs with Jack Daniel's barbeque sauce, or margaritas made with José Cuervo tequila.

Regardless of the type, strong brands must possess several key factors.[2] First and foremost, a brand must deliver a clear message quickly and succinctly. If the consumer is unclear about the type of product that is branded, or about the icon's link to the product, the brand will fail to generate positive identification for the user. A brand must also project credibility and quality. There are brands in the marketplace that have lost credibility; as a result, their value has severely diminished.

The leading brands elicit an emotional response from consumers. If a brand of food or a restaurant makes a customer's mouth water, the brand image is working. Such a Pavlovian response indicates that the customer links the brand with satisfaction in terms of a desired flavor or—at the optimal extreme—experience. These critical factors must converge to motivate action on the part of the consumer. For restaurants or food products, this action should take the form of initiating a purchase.

Keys to Successful Implementation

Strong brands usually equate to increased revenues. On-site operators have learned, however, that merely introducing a brand to an operation is not enough. The addition of one, two, or multiple branded concepts must be done to create an elevated image of the entire operation. Most eateries, for example, have at least one branded concept, albeit on a small scale. For instance, there is no reason why an operator cannot introduce a manufacturer's brand (such as Grey Poupon, cited earlier). Yet this doesn't necessarily communicate much to the customers.

On the other hand, an artful combination of national, regional, signature, and manufacturers' brands can convey to customers that a particular eatery is dedicated to quality (communicated by the familiar brand names and imagery) and that the offerings are expansive (illustrated by the multiple-brand approach). Sometimes referred to as **umbrella branding,** this integration of diverse yet complementary branded concepts and products ultimately serves to create a branded image for the eatery itself. This, in turn, results in greater customer loyalty, higher traffic flow, and improved positioning in the marketplace. Of greatest importance, it means greater revenue and higher profit.

An excellent example of umbrella branding in practice is illustrated in the best practice employed by the Hospital for Sick Children in Toronto, Canada.

BOOSTING REVENUES THROUGH BRANDING AT THE HOSPITAL FOR SICK CHILDREN (TORONTO, CANADA)

It might be argued that most children's hospitals experience greater participation in their cafeterias owing to the greater number of visitors. Some would say that foodservice operations in such institutions need not strive for high quality, given the greater patronage. The cafeteria in Toronto's Hospital for Sick Children looks at the greater patronage as an opportunity to maximize revenues and create a positive dining experience for those who are—in most cases—experiencing difficult life situations.

To this end, the foodservice management team has adopted branding in a unique way. Given limited capital funds, they identified brands such as The Donato Group's Made in Japan Teriyaki Experience (a well-known brand in Canada shown in Figure 4.4) as purveyors of products that their target market would enjoy. They then partnered with such companies, agreeing to pay a percentage of revenues for sales of their products and to buy key items (including paper goods and employee uniforms) directly from them. In exchange, the purveyors retrofitted existing space in the cafeteria to best profile their concepts.

Thus, the Made in Japan Teriyaki Experience, which leads in terms of sales of branded menu items, features dedicated equipment along with color-coordinated counters and neon signage. Other brands that include similarly adorned space are Pizza Pizza—a very popular pizza brand in Canada—Maple Leaf Foods' Bittner's Deli, and The Donato Group's Mrs. Vanellis (see Figure 4.5), an Italian concept.

Other brands that are featured but for which royalties are not required include Campbell's Soup, which is merchandised at a soup bar with eight different soups and two stews or chilies daily, and Coca-Cola, which is offered via a beverage wall. In addition, the foodservice department has created their own signature brands such as the Chef's Corner, which rivals the Teriyaki Experience in terms of daily sales, and The Grill, featuring items grilled to order. Signature brands on the horizon include iterations of a bread station and a salad bar.

The brands do more than expand choice and enhance appeal. They also drive profit, since each branded concept includes tested recipes, pricing models based on volume and serving sizes, portion-control methods, and training for staff. And, of course, the brands are appealing, particularly since the eatery now resembles a food court with a wide variety of menu choices.

If revenues are an accurate measure, the approach appears to be working. The department expects to beat the projected $2.7 million revenue number for the current year. Satisfaction is high as well. In fact, some of the products are so appealing that the foodservice department has integrated several of the brands with the patient menu in direct

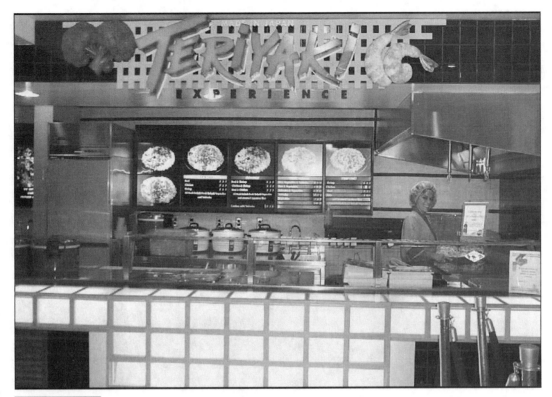

FIGURE 4.4 The Teriyaki Experience at the Hospital for Sick Children. Courtesy of the Hospital for Sick Children.

response to patients' requests. The individual pizzas from Pizza Pizza are particularly popular in the cafeteria and with patients.

The story doesn't stop there. The department has also utilized space outside the traditional dining areas and has outsourced the space to other operators. For example, Bento Nouveau, a sushi concept, has a kiosk. In return, the operator pays the hospital rent for the space and a fixed percentage of sales. Another brand is Mr. Sub; the operators cover build-out costs and operate on a multiyear contract.

With 372 beds, 2,955 FTEs, a number of patient-related services, and considerable square footage, the hospital has also outsourced more substantial areas to franchisees of national brands such as Burger King and Tim Hortons (one of North America's largest coffee and fresh-baked-goods chains, with some 2,000 stores across Canada). The outlets are located in different parts of the hospital and serve as tertiary outlets for employee and public dining. The agreements for these operations vary, but typically the operator is responsible for retrofitting and maintaining the space, and pays both rent and a percentage of sales for the right to operate within the hospital.

Altogether, the Hospital for Sick Children features more brands than are found in many food courts in large malls. The variety is considerable, and the quality is standardized through integration of proven concepts. In addition, the different outlets force the foodservice department to stay fresh and attuned to the same base of customers for whom all the outlets compete. At the end of the day, however, everyone benefits: The hospital enjoys stronger revenue streams, guests experience greater dining pleasure, and, it is hoped, all customers enjoy nurturing of both body and spirit through good food and service.

FIGURE 4.5 Pizza Pizza and Mrs. Vanellis at the Hospital for Sick Children. Courtesy of the Hospital for Sick Children.

Introducing multiple brands to create an enhanced identity for an on-site operation is challenging. For example, the introduction of brands such as McDonald's or Burger King may suggest to customers that the operator is neglecting the moral duty to serve only nutritious foods; this is particularly problematic in healthcare settings. The common joke that a hospital cafeteria serving several high-fat items does so to create customers for the healthcare organization is not greeted warmly by those associated with the host organization.

Other settings may not be as conducive to blatant integration of branding. For example, residents of some retirement communities want the perceived quality of branded items. Yet they do not want the visual vestiges of brands to enter what they consider their living environment. A case in point is a soda dispenser in a retirement dining facility. Customers want quality soda but do not want to see the Pepsi logo.

As alluded to earlier, a specific brand must complement the target market if it is destined to succeed. Introducing a Starbucks coffee may be appropriate in some settings but not others. By the same token, customers may have a strong affinity for a house-made pizza, savoring the "homemade" feel of the product, and may rebel against the introduction of a national brand.

Finally, branding cannot be introduced just for the sake of change. Branding is a complex process, and the introduction of any type of brand or brands is equally complex. Too many operators have been disappointed by after-the-fact analyses showing that the introduction of a brand did little to boost the bottom line. In poorly executed brand introductions, a branded product may cannibalize sales of other products with higher contribution margins. If the brand fails to communicate a positive message to the customer or fails to boost profit, then it probably was poorly selected or improperly introduced. This is an operator error and can be costly for everyone involved.

CHAPTER SUMMARY

Effective layout and design is the cornerstone to success for any on-site eatery. The first step in designing or renovating an operation is a two-stage planning process. Stage one is largely conceptual and provides the opportunity to categorize features. Stage two involves the development of the physical plan.

A feasibility study is typically conducted for larger projects to gauge whether a plan is reasonable, meets the needs of the target market, and is fiscally prudent. For smaller projects, decisions are usually based on common prioritized criteria. This often involves calculation of the IRR generated by the investment.

Several operating factors must be considered in planning. First and foremost is the accurate definition of the target market. The second is the development of a strong menu mix, one that is geared to the target market. Other related factors include the requisite skill level of employees and the amount and level of management at both the line and administrative levels.

Layout characteristics are a function of projected sales volume, menu complexity, and desired traffic flow. Related considerations regarding flexibility and the ability to serve numerous dayparts adequately while minimizing staffing requirements are important. Each of these is subject to space considerations and availability.

Regardless of the layout type, a marketing strategy involving the design of an adequate marketing mix to reach the target market is critical to long-term financial success. Merchandising is the other critical aspect of marketing. Branding, too, is a powerful tool for on-site operators. Branding, simply stated, is about image—both the image of the operation and the image with which the customer identifies.

KEY TERMS

feasibility study	marketing	national concepts
target market	marketing strategy	regional concepts
internal rate of return	marketing mix	signature brands
menu mix	merchandising	manufacturers' brands
scatter system	branding	umbrella branding
exhibition cooking		

1. Define a setting for a proposed eatery. Then describe the several considerations that are key during the planning phase.
2. Why is a feasibility study important?
3. Why is accurate determination of the target market important?
4. Assume that a small-scale renovation to a trayline-style servery is projected to cost $320,000. Using the projected increases to sales shown below, calculate the IRR. Assuming that the hurdle rate is 12.5 percent, should you proceed with the project?

   Year 1: $148,000
   Year 2: $137,000
   Year 3: $114,000

5. Define menu mix. Is a broad menu mix generally advantageous? Why or why not?
6. Which is better, a single serving line or a scatter system?
7. Outline the process for developing a marketing strategy.
8. Why is merchandising important? Is it possible for merchandising-related aspects of the operation to communicate negative messages to customers? Explain.
9. What lesson can we learn from the Crimson Catering example?
10. Define each of the branding types.
11. Why wouldn't the same branded concept be appropriate in all situations?
12. Discuss a situation wherein umbrella branding might be used to fully leverage the market position of an on-site foodservice outlet.

END NOTES

[1] A good textbook in this area is Almanza, B. A., Kotschevar, L. H., & Terrell, M. E. (2000). *Foodservice layout, design, and equipment planning.* Upper Saddle River, NJ: Prentice Hall. For more on the process, although much of the discussion applies to hotels, see Rutes, W. A., Penner, R. H., & Adams, L. (2001). *Hotel design, planning, and development.* New York: W. W. Norton.

[2] The literature on branding, and specifically on what is most important, is fairly extensive. Some good sources include Ries, A., & Ries, L. (2002). *The laws of branding.* New York: HarperCollins; Frankel, R. (2000). *The revenge of brand X.* New York: Frankel & Anderson; and Travis, D., & Branson, R. (2000). *Emotional branding: How successful brands gain the irrational edge.* Roseville, CA: Prima Publishing.

CHAPTER 5

PATIENT SERVICES

The process of taking a food order from a patient was once designed, it would seem, to be as impersonal as possible. The patient circled choices (assuming there were choices) on a paper menu for the next several meals. What was later delivered sometimes reflected these choices. Even when accuracy was high, the patient often no longer desired what had sounded good 24 hours earlier. Patients, after all, endure many physical, sensory, and emotional changes during a typical stay in a healthcare setting.

Service delivery was almost as impersonal. The aide arrived, placed the food on the rollaway table (often moving the few personal items allowed in the room), uncovered the hot dish, and left. In some instances, the patient couldn't reach the food and had to wait for a nurse to assist. In others, the patient wasn't even awake.

In order to discuss changes that have consigned such situations to the past, this chapter begins with a look at another major change—the shortened length of stay. Today, mothers delivering babies sometimes stay in a hospital for less than 24 hours. This is a striking change from the typical length of stay of the past, which ranged from three to six days. The chapter then explores how such changes impact foodservice.

Next, we discuss different cutting-edge approaches to service delivery. On-site foodservice managers now draw appropriately from practices in other segments to refine service delivery in the healthcare environment. To this end, the chapter looks at two distinctly new ways of delivering service, one akin to restaurant-style service and another to hotel-style room service, with each profiled in this chapter's best practices. To complete the discussion, the chapter explores related issues that impact patient services and that operators must address in order to stay ahead of the curve.

LENGTH OF STAY, CLIENTELE, AND FOODSERVICE

Healthcare providers have been classified historically as either short- or long-term care centers. The traditional cutoff was 30 days. Thus, a short-term stay hospital is one that averages less than 30 days. This was a logical approach when the average **length of stay** (days in the hospital) was somewhere between 7 and 14 days.

The national average today is far less than 7 days. Indeed, most healthcare providers are pushing for more outpatient activity, and HMOs are reclassifying many procedures as not requiring overnight stays. This means that customers—that is, patients—will likely leave the hospital to complete the recovery process at home.

This has enormous ramifications for the associated support services. For one thing, the throughput of patients is much greater. For example, consider a 500-bed hospital with an average **census** (an occupancy ratio calculated by dividing the number of beds "sold" or occupied by the number available) of 75 percent. If the average length of stay is 7 days, then the number of individuals using the hospital during a given year is about 19,554. If the length of stay is 3.5 days, however, the number of individuals doubles to 39,108. Thus, the number of **patient days** (the number of beds occupied during the period) is the same, but the number of individuals entering and leaving is much greater.

Now, consider that the average length of stay in the United States is approximately 2.82 days. This is shorter than at any other time in recent history. To make matters worse, dietary requirements typically change three times during this remarkably short length of stay.

From the foodservice provider's perspective, this means processing more information and at a quicker pace. It has many more specific implications as well. For example, initial contact with the patient is paramount. Second, the number of meals served will likely be fewer. And finally, menu offerings need to be retooled.

Making Initial Contact

When the average length of stay was longer, patients typically received a "standard" first meal such as a turkey sandwich. It was an innocuous selection, one that appealed to most people. With the standard issue, the patient also received a series of menus for the next several meal periods. The first meal was understandable and acceptable. With a reduced length of stay and the expectation that information should flow more quickly, patients are less likely to accept a standard meal, particularly if its purpose is to make the caregiver's job easier. Rather, patients view every aspect of their short stay with considerable scrutiny, often made more intense by the severity of their illness.

Thus, most good foodservice managers ensure that patients are greeted promptly and that diet restrictions and preferences are quickly integrated into whatever data-processing technology is used. Some larger institutions still use a standard meal but offer a replacement if the item is not satisfactory. Unfortunately, this tactic is inefficient, potentially doubles the cost of the meal, and does nothing to enhance the patient's experience.

Average Meals Served

With longer lengths of stay, patients often received on average almost three meals per day. Most often, they missed a meal during the time they were in surgery or when they were experiencing some type of procedure. As they approached a healthier state, they resumed a more regular eating pattern.

When the length of stay is only a few days, however, most people do not eat the equivalent of three meals per day. As discussed earlier, they are discharged more often in a condition that precludes normal eating activity. Thus, some acute-care centers find that the average number of **meals per day** is closer to two than the traditional three.

While this equates to a lower cost per patient day for meal service, it also motivates the foodservice department to respond to patient preferences and to achieve accuracy in terms of item delivery. If patients receive a total of only five meals during hospital stays, they expect each of them to be palatable. Patients also expect the items to correspond to their requests. With few meals delivered, the patient has less information on which to rate the department. Thus, every meal is critical in terms of meeting—and ideally exceeding—customers' expectations.

Menu Offerings

Traditionally, foodservice managers relied on one of two types of menus. The first is a traditional cycle menu, the type used in many on-site operations. Entrées are different each day for a set number of days (usually not seven, to avoid offering commonly recurring items on the same day of the week, also known as the "meat loaf on Monday" syndrome), and side dishes are designed to complement the primary items. Patients generally are allowed to choose a hot entrée, a cold entrée, or a sandwich. Typically with this approach, cafeteria and patient offerings are similar, since both follow the same cycle to facilitate food production.

The second type is a restaurant-style menu. Patients have the same menu each day, but it offers many choices. In a situation where the length of stay was 10 days, the menu designer ensured that a patient could conceivably order a different dinner entrée every day during the stay. Such an approach required careful production planning, since attempts to meet poorly determined forecasts could result in overproduction of some items (thereby wasting money) and underproduction of others (resulting in unwanted substitutions). While the cycle menu was the most common for many years, the restaurant-style menu gained popularity in many sectors, given the greater perceived control it gave the patient. The similarity to a restaurant format was also attractive from a marketing perspective.

With shorter lengths of stay, the restaurant menu is far more desirable since the number of offerings need not be extensive. In addition, the economies gained from a cycle menu are fewer when the cycle is shorter. The only downside to the restaurant menu in a shorter-length-of-stay situation is that the variety is less impressive. This is unfortunate since the large number of choices was one of the key attributes of such an approach. Nonetheless, it is not cost efficient to offer 11 entrée choices when the patient will, on average be having dinner for only two nights.

The shorter average length of stay and the resulting changes in operations are important for several reasons. Most importantly, foodservice managers must now rethink how the entire system operates in order to maximize satisfaction. The next section highlights two key variations that are quite valuable.

COMPARING RESTAURANT AND HOTEL ROOM-SERVICE APPROACHES

Two approaches have emerged to address patient dissatisfaction with food-related service. While both approaches eliminate the standard offering and often the institutional-style printed menu, they differ significantly in terms of logistics and system design.

The first approach is quite similar to traditional restaurant service, whereby a foodservice representative takes the order verbally and returns a short time later with the requested items. The second resembles traditional hotel-style room service, complete with room-service menus and a wait staff.

Restaurant Waiters in Hospitals

Imagine that you are a patient in a hospital. As mealtime approaches, you look for the paper menu in the materials you were given upon being admitted. You are surprised by a knock at the door. It is a foodserver who looks much like a waiter in a restaurant. In this case, however, the person is an employee of the hospital's foodservice department. He asks about your preferences, already knowing about your doctor-ordered dietary restrictions. (The waiter explains that the information is held in the small computer he carries with him.) You place your order, and a short time later the food arrives. When you order your next meal, the foodserver, a different waiter but one who displays the same level of customer orientation, knows which preferences you related to the server during the last meal. Does this sound too good to be true? It is reality.

Leading foodservice managers have taken the patient-focused approach discussed in Chapter 2 to a new level. They have merged traditional **restaurant-style service** with large-scale foodservice production, producing a marriage of convenience, fiscal conservatism, and conscience. The effect on satisfaction is obvious; not so apparent is the savings on food costs.

Taking meal orders closer to the time of meal service means that patients are less likely to change their minds about what they want. It also reduces the risk of changes in medications or diet that would alter food choices. Finally, it communicates to the patient that food will be delivered soon, eliminating the surprise element and the resulting need for food to be delivered later, when the patient is ready to eat. In concert, these results dramatically reduce tray costs by eliminating extra tray production.

Reductions in labor cost are also possible with this approach. With new technology that facilitates seamless data transfer, the data entered at the point of service are instantly available for dietitians to evaluate. In addition, once a database contains the information, it can automatically be massaged into production sheets for the kitchen.

Another noteworthy benefit, not directly apparent on the financial statement, is that patients feel empowered. Human beings are asking them what they want; they are offering service. Given that food delivery and visitors are the two most anticipated events in the hospital, this type of service meets a variety of intrinsic needs, even if the patient does not realize the immediate importance of having these needs met.

For an example of how this type of service is delivered at the highest level, see the description of the New Hanover Regional Medical Center room-service program.

BEST PRACTICES IN ACTION

PERSONAL SERVICE FOR PATIENTS AT NEW HANOVER REGIONAL MEDICAL CENTER (WILMINGTON, NORTH CAROLINA)

At one time, foodservice at the New Hanover Regional Medical Center in Wilmington, North Carolina, typified what one would expect to find at a large public hospital. Patients were given a standard meal upon admittance. On the tray was a paper menu for the next day, and patients were expected to complete the menu for the next several meals. Of course, they didn't know what restrictions might apply to their choices and often didn't even know if they would be able to eat that far into the future.

Along with this standard approach, meals sent to patients frequently included items they didn't want. The reason, simply stated, was that the dietary restriction had changed since the patient had placed his or her order. The only recourse was for the patient to contact a nurse; that person, in turn, would talk to the people in the kitchen and try to get an alternate meal delivered. With such a system, patient satisfaction was low and costs due to repeated meal delivery were extremely high.

The foodservice management team examined the problem using a customer-satisfaction approach. What were customers accustomed to when ordering meals? Furthermore, the group explored the steps involved in the process. In its original form, the patient completed a paper menu. The menu was then delivered to the diet office via the returned tray for analysis. Changes were made when patients asked for items that did not meet their prescribed diets. The diet office staff then collated the items and forwarded the information to the production staff, which in turn began preparing the appropriate food items. Errors were so numerous, however, that the cooks were accustomed to overproducing most main items since they knew they would need to have the food available when patients requested a second tray with different food choices.

Thus, the entire process boiled down to two key issues. First, the department needed to greet, inform, and elicit food choices from patients differently. Second, there needed to be a better—and

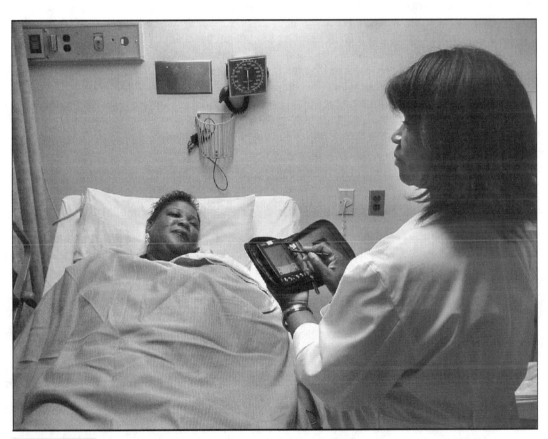

FIGURE 5.1 A Room-Service Attendant Takes a Patient's Meal Order at New Hanover Regional Medical Center. Copyright by New Hanover Regional Medical Center.

more accurate—way to massage the information so that production could be optimized.

The solution to both issues involved new technology. Developed by CBORD, a worldwide provider of food and nutrition services software and systems, a unique combination of hardware and software made the process more personable, efficient, and economical.

Today, foodservice representatives, as shown in Figure 5.1, sporting a handheld computer (or "palmtop") weighing less than seven ounces, visit each patient in any of the hospital's 600 beds. The computer is already programmed with the compliance information pertaining to the patient's diet. The employee then converses with the patient about what he or she would like for the next meal. The computer flags the items that are most popular and automatically deletes those for which the patient has expressed dislike. If a patient wants an item that is not in the computer, most likely because it does not qualify under specific dietary restrictions, the order taker can still place the order but can explain that the item must be modified accordingly or, in extreme circumstances, will have to be replaced until such dietary restrictions are lifted.

The data are then automatically uploaded from the palmtop to the diet office. Using specialized software, a single employee quickly reviews orders. Anomalies and special requests are flagged by the system for review by the dietitian. From the information, production sheets are automatically generated, replete with information on quantities, special instructions, and so on. This is accomplished through minimal manual data entry, thereby reducing the opportunity for error and greatly increasing efficiency.

Has the solution worked? The answer is evident in patient-satisfaction scores and cost savings. In the short time that the full system has been in place, the feedback from patients has been overwhelmingly positive. They experience foodservice in a familiar and comfortable way, and they perceive greater control over what they eat.

The cost savings are equally impressive. Food cost has been reduced by the drastic reduction in tray duplication. Moreover, labor cost has been reduced in terms of tray delivery and diet office personnel.

As for the cost of the technology, the foodservice managers project a positive return on investment in less than two years. But the main benefit—one that they cannot quantify as easily but feel is more important—is that patients are happier, and, they hope, are able to recover from their illnesses more rapidly as a result.

Hotel Room Service Without the Hotel

While the restaurant approach is laudable, it is not the only way to reshape patient foodservice. Despite the personal touch that is so impressive when restaurant-style service is used, meal periods are still narrowly defined. This is where the application in healthcare diverges from that traditionally seen in restaurants. After all, customers in restaurants dictate when the waiter will greet them by timing their arrival at the restaurant. If they want an early lunch or an after-movie dinner, they simply go to the restaurant at the corresponding time. In healthcare, the waiter arrives shortly before the traditional serving time for the corresponding daypart.

This is problematic for patients who want breakfast at noon or dinner late at night. Hotels found the solution to this problem long ago and aptly named it **room service.** At almost any full-service hotel, guests can pick up the phone and order anything from the in-room menu. A short time later, the food arrives.

Such an approach is amazingly well suited to a healthcare setting in which patients have a variety of diagnoses and therapies that often take hours to complete and span traditional mealtimes. In addition, shorter lengths of stay often mean that patients have less time to adapt to unfamiliar meal schedules; just think of the poor patient who has worked nights for 20 years and is expected to rise at 7:00 A.M. for breakfast! Room service offers the flexibility that has always been lacking.

The other major attribute from the patient's perspective is power and control. In a hospital, patients often are told when they can sleep, what pills they need to take, and even when to use the bathroom. Having a choice of when (and what) to eat sometimes represents the last vestige of personal control and empowerment for a patient.

While the notion of offering room service in healthcare is not completely new, finding solutions to the accompanying challenges is. For example, the order-taking process has always been a hurdle. A common phone number, accessed from the patient's in-room phone, is simple enough as a first contact step. On the receiving end, however, the problem has always been staffing and data management. Even if a diet clerk answered the phone, how would he or she know what diet restrictions were in place for that specific patient? Emerging technology is simplifying things. Using a single data terminal, or even two terminals if all information isn't kept within a single system, foodservice employees can easily take a patient's food order. In its most advanced application, a single computer, which is tied to the healthcare organization's main system, integrates doctor-ordered dietary restrictions with information on patient preferences available from tertiary sources. In real time, the order taker can help patients order items while adhering to their doctors' orders and also make suggestions (e.g., "Would you like a milkshake with that sandwich?"). If the process is optimally mapped, it can even prompt the order taker to ask about condiments and side dishes.

The other challenge that has kept room service on the foodservice manager's wish list, especially when the implied goal is to prepare most items to order, as in higher-end hotels, pertains to production. With advanced food-preparation technology (often integrating some of the newer production techniques discussed in Chapter 3), unpredictable demands are now more easily managed. These same approaches also ease the challenge of offering breakfast, lunch, and dinner items throughout the day.

It is readily apparent that this style of service would increase satisfaction by increasing patients' sense of control while accommodating many diverse dining preferences. Moreover, in understanding the related technology, the execution seems reasonable. The remaining challenge, then, is cost.

Undoubtedly, labor costs are higher with this style of service than with traditional, impersonal foodservice. Surprisingly, the reduction in food cost typically offsets such increases. This is due to two main functions of the system. First, there is no redundancy in terms of multiple tray deliveries. Patients order when they're hungry and, knowing the meal is forthcoming, are ready to eat when the food arrives. Changes to diet orders can be tracked more closely; the patient can also be guided to order accordingly, making later substitutions unnecessary. Finally, patients call for meals or snacks only when they want them, which is typically less often than the traditional three meals a day, which are sent in the old-style system whether they are wanted or not.

There are many nuances to room-style service, and different patient mixes demand certain special accommodations. Nonetheless, this is another wonderful method of breaking away from doing things as they've always been done. One healthcare setting that offers particularly impressive room service is Children's Hospital of Cincinnati.

ROOM-SERVICE DINING AT CHILDREN'S HOSPITAL (CINCINNATI, OHIO)

Many people are accustomed to lying in bed in a hotel and ordering room service. All that is required is a phone call and a short wait for the server to deliver the food. We take for granted that the food will be delivered in a timely manner, and we know that it doesn't really matter if we order waffles for dinner or a roast beef sandwich at 10:30 A.M. It's what we want, and we expect to get it.

FIGURE 5.2 Children's Hospital of Cincinnati's Room Service Menu—Entrées and Side Orders. Copyright by Sodexho USA Children's Hospital of Cincinnati.

Patients at Children's Hospital of Cincinnati enjoy the same type of service, even ordering from a menu that looks very much like a hotel room-service menu. (See the menu in Figure 5.2.) At any time from 6:30 A.M. to 11:00 P.M., a patient may dial "F-O-O-D" and will be cheerfully greeted by a diet clerk. Each clerk is equipped with two computers, one linked to the hospital's computer network, which has up-to-the-minute information on each patient, and another that is similar to a point-of-sale system in a restaurant, complete with touch-screen interface and full menu-item descriptions. In real time, the clerk can cross-reference patient data, largely to ensure that patients with doctor-prescribed restricted diets receive appropriately prepared food items.

While talking with the patient, the clerk enters the food order into the point-of-sale system. The order is automatically fed to a series of thermal printers in the kitchen. This process mirrors the system in most restaurants; orders for cold food items are routed to the cold-food prep area, orders for grilled items to the grill station, and so on. A copy of the complete order is also printed in the expeditor's station; this employee assembles the complete meal and queues it for delivery in a temperature-controlled cart. And even though the hospital consists of three separate buildings, the foodservice department at Children's Hospital of Cincinnati is able to tout a 55-minute service promise from the time of the call.

The implications of this room-service program are impressive. First and foremost, any patient in any one of the hospital's 330 beds can order anything he or she wants at any time. Thus, patients receive only what they want and only when they want it. Second, there is no tray duplication because of misinformation regarding room or diet changes. Third, there is no need for preselected meals, thereby giving patients a sense of comfort and control. Finally, management can focus on whether the meal itself is pleasing, as opposed to worrying about whether patients have received what they wanted—or anything at all.

Virtually all of the concerns that were first voiced by different constituents of the healthcare provider have been alleviated since the system was introduced. Some thought that patients would abuse the system, ordering more than the typical number of meals per day. In actuality, the foodservice department now produces fewer meals than they did with the traditional tray service, owing largely to the enhanced accuracy and elimination of tray duplication. Others were worried about patients ordering foods that aren't necessarily the healthiest items served in hospitals. Evidence indicates that patients, particularly children, experience improved recovery when the food and the environment are more akin to their natural surroundings. On that topic, some of the most popular items include pizza, chicken tenders, and chicken noodle soup.

Some administrators were concerned about the cost of labor. After all, the system requires more diet clerks, who field the room-service calls, and more production staff, who must be on duty for the extended serving times. The savings in food cost, again realized through the elimination of tray duplication, reduction in overall tray cost (since patients order only what they want instead of receiving standard meal items), and increased accuracy in goods produced versus goods ordered, offset the increase in labor cost proportionately. Thus, room-service dining was launched at a cost equal to that of the old system.

While the concern over cost is moot, the biggest questions pertain to quality and guest satisfaction. What do the patients think? Prior to the launch of the room-service program, Children's Hospital was in the 10th percentile in overall satisfaction with food and related services (including temperature of menu items, selection, and service, among other variables) as compared with similar institutions. Following the implementation of the room-service program, the satisfaction scores rose astoundingly, putting the organization in the 99th percentile in terms of satisfaction with foodservice! (The ranking is based on data provided by Press Ganey Associates, Inc., a third-party firm that collects patient-specific data for healthcare providers and maintains a national database of patient satisfaction information.) These extraordinary results might even persuade hotels to upgrade their room-service programs.

RELATED IDEAS

Obviously, there are other issues that we could discuss pertaining to patient services that affect satisfaction, cost, and the overall impact on the foodservice department. It is difficult to separate some aspects of patient service from others, because they are all interlinked. In fact, it is the resulting package that defines the experience for patients. Nonetheless, other issues include those associated with point-of-service materials, training of staff for serving the customer, matching menu items with the target market, and, last but not least, offering added services as a means of enhancing both perceptions and revenue.

Point-of-Service Materials

First impressions are important whether on a date or in the hospital. Yet patients are often introduced to the foodservice team by dog-eared merchandising materials, dirty menus (when a room-service program is in place), or table tents that need replacing. The implied message of such point-of-service materials is clear: "We don't care about how you perceive us." Why should patients think the department would take care in preparing the meals if it doesn't take care in merchandising itself?

Any program, whether restaurant-style, hotel-room-service-style, or any other, deserves to be marketed. Patients, like most customers, respond positively to consistent communication; more often than not, they also embrace attractive imaging as part of the message. Hence, marketing efforts can quickly get torpedoed if poorly produced or inappropriately maintained materials are used.

This also means that any patient-communication material, including menus, tray mats, tray tents, and table tents, should be developed as part of a comprehensive marketing strategy. Color schemes must be complementary, images must be reflective of social identities in the target market, and messages must be written clearly and succinctly. The number of fonts should be limited; background imagery should be thoughtful and should reflect the context of the application.

Simply put, materials should be patient-friendly. This sometimes means printing in two or more languages. It also means that complex messages need to be reduced to easily understood, focused lists. Fewer words are always better.

Training of Service Staff

Just as marketing and merchandising are important, so are the perceived attitudes of foodservice representatives. Any restaurant patron knows that good food served with a bad attitude doesn't go down as well as when it is served in a pleasant, congenial manner. The same holds true in the hospital. In fact, service is perhaps even more important since the food may be less than optimal due to physician-prescribed dietary restrictions.

The patient has just learned that more surgery is needed. It has been several days since he could eat anything but broth, and he is told that he cannot have any solid food for another 48 hours. How important is the attitude of the person who brings his "meal"?

This is a matter of training. Line employees who deal with patients are sometimes the lowest-paid individuals in the foodservice department. They may not have good skills, but they can be taught many things—and most are more than willing to learn. It falls to management to ensure that such training takes place. Moreover, the message that service counts must be reinforced constantly.

Does the Menu Match the Target Market?

There was once a hospital in Philadelphia that had a well-defined patient mix, partly owing to the fact that it serviced a narrow corridor within the inner city. Most patients were of Hispanic descent. The age distribution was somewhat bimodal, with one group around 38 years of age and another around 68. Most patients came from homes in which the food reflected the ethnic background of the inhabitants.

The hospital offered good service; by most accounts the food was served in a timely fashion, with hot food arriving hot and cold food served accordingly. There was even a high level of accuracy in terms of meals matching what was ordered.

The problem, obvious to patients but not to the administration, was that the menu items did not reflect the patient population. Meatloaf with mashed potatoes, sautéed chicken with mushrooms, and a vegetarian ratatouille with subtle cabernet overtones were some of the dishes that the foodservice department proudly featured. Evidence generated from a survey conducted by an outside consultant revealed with alarming clarity that patients found the food bland, uninspired, and—in some cases—unpalatable.

What patients said they wanted is what patients in all parts of the world say: **comfort food,** food that is nurturing, that satisfies the soul, and that evokes feelings of warmth and well-being. The challenge is that different people have their own definitions of what food they find most comforting. The menu at the hospital was clearly what the management felt was comfort food. Unfortunately, they failed to view the issue from the patients' perspective.

Additional Offerings—for a Price

Every on-site operator knows the worth of offering specials during a meal. For instance, many operations in New England, including those in B&I, healthcare, and education, offer the traditional lobster roll periodically; operations in Southern California may offer fish tacos. The purpose of such offerings is to (1) increase revenue, (2) increase profit, or (3) increase customer satisfaction. The best specials accomplish all three goals.

Surprisingly, patient services rarely employ a similar strategy. The argument is that such an offering would be too difficult, given the disparity of different diets, and that it wouldn't be cost effective. Some operators have found that such thinking is unnecessarily prohibitive.

For years, many patients have elected not to share a room, even though this decision costs money. When available, some patients—as many forward-thinking foodservice managers can attest—will also pay extra for special meals. By coordinating their efforts with billing and other related departments, foodservice can easily offer menu items that are too costly to offer systemwide but that may offer extra appeal—enough so that patients will pay the extra money. This increase covers the additional raw product cost, the labor needed to prepare the item, and a reasonable profit.

This is also a wonderful idea when paired with menus for visitors. Many times the spouses, parents, and children of patients want to enjoy a meal with their loved one and are more than willing to pay the market price for the opportunity. The need is already in place; finding a way to execute a profitable response to this need is just good business.

CHAPTER SUMMARY

The average length of stay in most healthcare institutions is shorter than ever before, resulting in a greater average throughput of patients. This situation indicates, for the foodservice department, that making initial contact and delivering optimal service is critically important. The initial contact should serve to formally introduce the patient to the foodservice operation, to collect information regarding dining preferences, and to facilitate order taking for the first meal. The goal is to avoid a default or standard-issue first meal.

Restaurant-style service represents a powerful departure from the conventional impersonal paper menu system. With the restaurant style, patients can place their orders closer to corresponding mealtimes. This means less chance of dietary orders changing between ordering and delivery. And that means fewer mistakes, with the resulting need for replacement meals. The end result is lower systemwide food cost.

Hotel-style room service is another approach that enhances patient satisfaction, perhaps to an even greater extent. Patients place their food orders whenever they want. Through the use of new technology, the foodservice representative can help with substitutions resulting from special dietary restrictions while offering personal service. The labor cost with such an approach is higher, but the offsetting food cost reductions are realized because delivery of meals is accurate and patients order only what they want. Both approaches serve to maximize patient satisfaction.

KEY TERMS

length of stay	meals per day	room service
census	restaurant-style service	comfort food
patient days		

REVIEW AND DISCUSSION QUESTIONS

1. Why is the old way of providing foodservice to patients less than optimal?
2. Name the most important ways in which the reduced length of stay affects the foodservice department.
3. As a patient, would you prefer to see a cycle menu or a restaurant menu? Why?

4. Of the two approaches described, which is more likely to offer a higher level of customer satisfaction?

5. In terms of restaurant-style service, what type of uniform should hospital waiters wear?

6. Why has it taken so long to integrate the hotel room-service approach in healthcare?

7. What message do patients receive from point-of-service materials that are dirty and worn? Why is this important?

8. What is a viable approach to assess whether a menu matches the dietary preferences of an organization's target market?

9. A foodservice manager has decided to launch a special menu for visitors to the hospital who want to dine in the room of their loved one. What are the chief concerns and operational challenges that should be considered?

CHAPTER 6

CATERING AND SPECIAL FUNCTIONS

The enterprise of catering reflects the times. Yet, even before the refinement and dissemination of catering expertise was formalized, a host of novel practices distinguished this specialized segment of the foodservice industry. Some of these approaches were so novel, so extraordinary, that they still are considered remarkable.

Take, for example, preparations made for gatherings hosted by the Honorable Edward Russell, captain general of the English forces at the height of William III's reign in the late 1600s. With particularly impressive innovation, the caterers used the fountain in his garden as a giant punch bowl for mixing his libation of choice. The recipe included 560 gallons of brandy, 1,300 pounds of sugar, 25,000 lemons, 20 gallons of lime juice, and 5 pounds of nutmeg. The bartender rowed about in a small boat, filling up punch cups for the awed guests. Another example dates back to the early years of that same century, during a time when the wealthy English were known for "surprise" pie. This odd culinary creation was a main dish, and was brought to the banquet table by the catering staff with great fanfare. It was opened ceremoniously, sometimes by the host. Upon the first cut, out of the pie leaped all sorts of live creatures: frogs, rabbits, terriers, foxes, and, as the nursery rhyme claims, four-and-twenty blackbirds. The size and number of animals came to be associated with the sophistication of the catering efforts.

With the effects of such extravagance on dazzled guests as a backdrop for a discussion of catering, this chapter addresses the potential impact of the catering function for on-site foodservice operators. First, it considers the importance of artfully delivering catering services within the host institution. Next, it looks at off-site catering as a venue for untapped revenue—and profit. On-site operators in almost every segment of the industry have realized this and are creating remarkable dining experiences using the resources at hand. The work of Crimson Catering briefly discussed in Chapter 4 is a good example from the college-dining segment; this chapter's first best practice provides an even more provocative example. Finally, the chapter considers the complex issue of catering-menu pricing. A variety of methods exist, from the overly simple to those that include objective and subjective variables.

IN-HOUSE CATERING: ADDING VALUE

Catering can be approached from a variety of perspectives and integrated with an on-site foodservice operation for a number of reasons. One typical approach is to introduce catering to build customer awareness within the host organization regarding the foodservice department's ability. The integration of catering in such a case is typically accomplished by using existing staff, cross-training employees for service, and tapping the talent of the production crew—thereby eliminating the need to add dedicated catering FTEs.

Regardless of the reason, adding catering to an on-site operator's repertoire of services adds value to the host organization. Quite often, on-site operators can perform in-house catering more cheaply and efficiently than any outside vendor. The facilities are already nearby; the billing process is simple; and the relationships necessary to promote the business quickly are already in place.

While the advantages of adding value through catering cannot be dismissed, a number of concerns are paramount for success. These include available facilities. For example, does the on-site foodservice provider have the necessary space and equipment? An associated concern pertains to employees. Do the current staff members have the talent to pull off diverse catered events? What about customers? What are they expecting? What are they used to? These questions are often answered by examining who has provided similar services in the past. Finally, how should the cost of catering be transferred internally within the host organization? In most instances, the payment is simply an internal transfer of funds. As such, the pricing may be intended to cover only the related variable costs. Nonetheless, careful analyses are required to ensure that such pricing is accurate and that all related variable costs are covered. For operations that price catering on a market scale, the last major section in this chapter offers the necessary information.

Facilities

For an on-site foodservice department looking to introduce or grow its in-house catering service, the facilities can serve as an asset or a liability. Front-of-the-house areas are not much of a concern since most host organizations have meeting rooms, open office areas, or similar space in which food and related services can be enjoyed, at least for small- and medium-sized functions. In addition, there is always the cafeteria, which can be utilized between or after traditional serving times; a specific eatery, in cases where multiple outlets are used, can even be closed for a special meeting or function.

The real bottleneck, when one exists, is the back of the house. Is there enough workspace to prepare meals for both normal operations and catered activities? On a related note, is there adequate refrigerated and dry-goods storage space for both enterprises? These questions cannot be answered without empirical analysis. Thus, objective determinations are required before accepting responsibility for new or additional catering services.

Employees

While storage can be adapted for unusual events—for instance, underutilized areas can be used to store various items temporarily, or refrigerated trucks can be used for temporary storage of prepared food items—employees' abilities are not as malleable. To be sure, employees can be trained to do a variety of tasks. Talent sophisticated enough to produce exquisite catered food is not, however, something that is learned by skimming an Iron Chef compendium.

While production talent is the key concern, a related matter is the service staff. Can staff members who have been trained to work in an old-style B&I cafeteria tray line be expected to serve the board of directors a gourmet meal? Moreover, is the labor—assuming that employees have the requisite skills—available to work the flexible hours?

On the flip side, catering can add sizzle to jobs that employees view as mundane. At the most basic level, an employee might enjoy delivering a tray of breakfast croissants to the executive boardroom, taking responsibility for setting up and breaking down the food table. At the other end of the spectrum, the opportunities associated with upscale events such as bartending, waiting tables, or serving can be provocative. Thus, in-house catering can do more than just add value to the host organization; it can also provide an exciting opportunity for foodservice employees.

Customers and Competition

On the one hand, it is fairly easy to develop a profile of in-house customers for an on-site catering venture. After all, the customers routinely visit the employee eating area. The downside is that even with this knowledge, the foodservice staff must be able to understand the catering needs of the market.

Sometimes this is a matter of tracing requests for special items over time. Did the purchasing department ask foodservice to send up box lunches during their strategic planning meeting last year? Has the human-resources department historically asked for pastries during the monthly recruiting fairs? Such historical data serve as a beginning for matching the target market with an appropriate range of catering services.

It is important to realize that when a foodservice operator takes on the role of in-house caterer, he or she is implicitly stating to the other departments of the host organization that the foodservice team is capable of planning menus, decorating meeting spaces accordingly, staffing the event in terms of services related to food (including replenishing items), serving guests, and cleaning the area after the event concludes.

The basis of comparison will be other caterers in the area. Customers, even those with limited experience with caterers, will compare everything the on-site provider does with catered events they have witnessed in the past, particularly those that left positive impressions. These expectations can sometimes be quite substantial. Customers will similarly expect the in-house caterer to do everything they expect of outside vendors, including catering office parties, cocktail parties, open houses, small dinners for executive breakouts, retirement parties, and employee parties.

Thus, the on-site provider must also know the competition through a process referred to as **environmental scanning.** This entails answering a number of questions. For example, what are their capabilities? What is their pricing strategy? Do they have a good reputation? What are their primary strengths and weaknesses? Finally, are there any ties within the host firm (for example, the owner of a local caterer is the brother-in-law of the host organization's chief executive officer)?

Awareness and analysis of the competition is an exercise worth reviewing at regular intervals. Just because the foodservice operator is unaware of a new caterer—possibly a retired chef from the local four-star hotel—does not mean that customers will share this ignorance.

Internal Charges

As alluded to earlier, pricing is a critical component of catering, perhaps more so for the in-house foodservice provider. In many instances, the host organization expects the foodservice operator to charge only for the variable expenses associated with the catered event. For example, suppose that the host organization's administration orders 100 box lunches for a corporate retreat. And suppose that these cost the foodservice provider $3.24 to prepare and that this preparation was done using the staff already scheduled and on duty. The host organization may stipulate that the foodservice department will transfer the cost of only the $324 for the food, contending that the labor was not an additional variable expense. But what about the cost of paper goods? These variable costs must also be transferred to maintain equity.

When catering is priced this way, the foodservice department does not experience deterioration of its operating margins and it adds value to the host organization. No outside caterer could operate on such a basis. Moreover, the good will that such exchanges produce is immeasurable. The remaining challenge is to maintain managerial sanity while trying to deliver maximum value and service to the host organization, all the while knowing that the activities do little to build the bottom line.

OFF-SITE CATERING: VENUE FOR UNTAPPED PROFIT

While on-site catering brings value to the host organization in terms of convenience, service, shared synergies, and beneficial economies, off-site catering is a richer venue for profit. In addition, off-site catering affords the on-site foodservice operator the chance to showcase the talents of the staff, enhance the reputation of the department (through catering high-end, high-profile functions), and offer employees the opportunity to venture beyond their regular jobs.

There are a number of potential land mines, however, when the on-site operator leaves the protective walls of the host organization. Typically, these include menu development, sales and marketing, and project coordination. While these operational issues affect any business involved in catering, they are potentially more problematic for the on-site operator, who cannot lose sight of the core business while engaging in activities that offer considerable potential benefit.

Menu Development

A catering menu program, like any other menu-related issue, must be geared to the target market. The style of service, price range, and item variety define the catering efforts and speak either positively or negatively to potential customers. Thus, the message delivered in such material must be clear and must reflect the expectations of the intended customers.

Even if the reputation of the on-site foodservice operator is excellent, it will not necessarily follow the operator into the catering end of the market. A school or hospital may achieve national acclaim for its foodservice, yet this does not mean that a young couple would hire the on-site operator for their wedding on the basis of reputation alone. If the on-site foodservice manager has a reputation as a wonderful caterer, however, this may be reflected in the reputation of the on-site operation. This one-way relationship is driven by the unfortunate premise held by many customers that catering is difficult and complex but on-site foodservice is simple to execute.

The menu program, then, is the main vehicle in delivering this message. And, as discussed in Chapter 4, everything flows from the menu. To reiterate, the menu should reflect the ability of the operation, its equipment, and its talent. A creative and flexible menu allows the operator to showcase the talents of the department while catering to unique customer characteristics such as size and location. A set, inflexible menu may be simplest to execute, but it will too narrowly (and most likely inaccurately) meet the target market's needs.

Sales and Marketing

A convenient way to approach the challenges of sales and marketing for the on-site operator who offers off-site catering is to use the traditional marketing categorization known as the **four Ps**—product, price, promotion, and place. The menu is prominent in the marketing effort of any foodservice enterprise, as it represents the core product. This holds true for caterers as well. In fact, the menu is the primary promotional tool.

Some operators marry the menu with promotional materials, featuring pictures of past events, client profiles, and references. While these efforts may produce the desired effect, they still depend on the appeal of the menu to drive sales maximally. In addition, catering menus set the initial expectations for customers. If the menu is elaborate in design, the expectations are that the execution will be similarly detailed and intricate. While seemingly a good thing, meeting or exceeding such expectations can be difficult, particularly for small functions.

The extensive discussion on pricing in the following section underscores the importance of this aspect of the catering business from a marketing perspective. As noted earlier, many catering customers use price as the primary decision criterion. In addition, menu prices can indicate what part of the market the operator is targeting. Such a decision can quickly dictate failure if made incorrectly or without pertinent information.

The ideas of product and price lead directly to promotional efforts. On-site operators have an advantage in that they have a captive audience to whom they can market, using a product–price combination that they hope appeals to this audience. This enables narrowly targeted promotional efforts.

The downside, as noted, is that the positioning of the on-site operation may not reflect the type of catering the operator is promoting. This disconnect can confuse the promotional message. Similarly, the number of employees to whom catering would be of value may be limited. Nonetheless, this group often represents a good starting point. Word-of-mouth advertising from this group is also sometimes an effective vehicle for generating business.

Other promotional efforts include direct mailings, media advertisements, and point-of-sale advertising. These vary in efficacy based on the target market, market mix, and economic conditions. They also vary greatly in terms of cost. The best advice for promotional efforts, therefore, is to apply cautious spending that is in direct proportion to the projected level of business.

The last of the four Ps, place, is a tricky one for on-site operators. While some on-site operations have the luxury of meeting space that can be used for external functions, most do not. This means catering events at locations where equipment must be quickly integrated on a temporary basis—which can involve formidable logistical obstacles. The first concern is food safety. Can food products be transported and either served or cooked without sacrificing quality food-handling procedures? Everyone has read of a party that was remembered because the guests fell ill with food poisoning. This is not something with which a caterer wants to be associated.

The second issue is quality food production and service. Moving dinner plates from a commercial kitchen to an adjacent dining room raises many issues. The problems grow exponentially when the kitchen is temporary and equipment is at a premium. Too often customers expect banquet-style foodservice to be mediocre. The on-site operator who can cater such events and positively alter these expectations is one who can also maximize profit. An example of just such an operator is featured here as a best practice.

BEST PRACTICES IN ACTION

CATERING AS A COMPETITIVE DIFFERENTIATOR AT SOUTH NASSAU COMMUNITIES HOSPITAL (OCEANSIDE, NEW YORK)

Many on-site foodservice operations offer some level of catering. Some even compete with local commercial caterers for business throughout the community. In some B&I settings, catering represents a considerable contribution to the bottom line. Similarly, many school and university foodservice operations leverage the talent of their culinary personnel, and take advantage of having already covered the fixed costs, by catering to in-house constituents including administration, fraternities, and sororities. Only a handful of on-site operators, however, extend their prowess to such an extent that they are defined by their impressive abilities in this secondary venture.

The culinary team at South Nassau Communities Hospital in Oceanside, New York, is one of these rare exceptions. Of course, they provide quality foodservice to the patients in the 429-bed facility, as well as to the employees and guests who dine in the cafeteria. It is the catering, however, that sets the operation apart. Bringing life to events ranging from small dinner parties to galas of 400 or more people through upscale delivery of fine food and service is the pride of the department's certified executive chef and his assistants.

The foodservice operation boasts several advantages. First is a glassed-in solarium adjacent to the cafeteria that is directly accessed by the kitchen. This room, which is used for lunch-hour dining and special events, provides an attractive backdrop to the operation's haute cuisine. Another advantage is the herb garden, which provides much of the fresh, aromatic ingredients for catered meals during the temperate months.

This is not to say that things are easy for the healthcare foodservice provider. Given the constraints of the healthcare-business climate, as discussed in previous chapters, the foodservice management is allowed only a single dedicated FTE for catering. The labor required to staff special functions, therefore, must come from the on-site staff. This translates into a systemwide emphasis on cross-training that includes the majority of the other 71 FTEs.

The department also does not rely on traditional flashy catering brochures, which can be pricey to produce. Instead, it depends on word of mouth as the primary vehicle for advertising. The administration and the culinary staff agree that the best way to spread the news is to showcase the food served daily as examples of what the department can do. Of course, the kitchen does not routinely serve lobster fricassée or stuffed quail to patients and guests. The food in the cafeteria, however, often serves as a test market for potential catering menu items. Customers benefit in that they get to try the dishes firsthand and also provide immediate feedback on the various items.

Just how much catering does the operation do? Some of the functions are for in-house customers, a sample of which is shown in Figure 6.1. The price is transferred internally based only on the

FIGURE 6.1 A Sampling of South Nassau Communities Hospital's Catering Efforts. Courtesy of South Nassau Communities Hospital, Oceanside, New York.

cost of goods. External functions, however, are priced competitively. Assuming that all functions reflected market-based prices, revenues would approach $1 million a year. Granted, the department has developed their clientele and the associated sales over many years. Nonetheless, the popularity of the food and service has resulted in a 300 percent increase in catered sales over the past four years.

How good is it? Recently, the department catered an event for a national meeting on breast cancer. The centerpieces were so impressive and the food was so memorable that guests of the functions first asked who the caterers were (assuming that the organizers had hired one of the local upscale restaurants). The follow-up question was whether the department was available for weddings.

Project Coordination

In 1911, Fannie Merritt Farmer authored *Catering for Special Occasions*.[1] In describing the state of the catering industry, she offered the following appraisal:

> In these days of rapid transit, by sea as well as by land, the markets of the world are brought almost to our very doors, and we have a hundred combinations to our grandmother's one. We, therefore, receive our guests more formally; we make preparations for their coming, and take pleasure in giving them a meal which shall vary from the humdrum order of culinary production. The fashion in entertaining, as in so many other things, has changed, and consciously or unconsciously we conform to the new standards.

These words are just as true today as they were nearly a century ago. The difference, which was discussed at length in Chapter 2, is that customers' expectations contribute to the definitions of the "new standards." This makes the problems that were common decades ago even more challenging today. Fortunately, on-site foodservice managers have evolved at a pace that rivals—we hope—the complexity of the problems. Thus, the final issue for off-site catering is the complicated one of project coordination.

In its simplest form, a catering project involves filling a customer's order for food; this is true whether the service is on- or off-site. Taken a level up, it may involve scheduling a room and then delivering the food to that location. Up another level or two, the catering task may involve bidding for a large event, generating proposals and contracts, and then delivering what was promised. At all levels, food must be received and correctly inventoried, vendors paid, and customers contacted for service-quality analyses. The steps involved in managing the logistics alone require a sophisticated system.

For those on-site operators who cannot afford multiple staff members to deal with such tasks, coordination of the catering process requires considerable organizational skills. In addition, the labor scheduling, above and beyond that required for normal on-site operations, is problematic and requires constant attention. Finally, limited human capital also translates into a need to train effectively; it may also mean hiring those who possess the flexibility necessary to succeed in an operation wherein each day brings a unique combination of customers and challenges.

While management is the key to success in coordinating and executing all the tasks associated with catering, some operators have adopted approaches that make their lives easier. A fine example of just such an operator is the foodservice management team at Sunnybrook and Women's College Health Sciences Center in Toronto, Canada, which uses software to facilitate delivery of the best in catering services.

BEST PRACTICES IN ACTION

USING A SYSTEMS APPROACH TO CATERING AT SUNNYBROOK AND WOMEN'S COLLEGE HEALTH SCIENCES CENTER (TORONTO, CANADA)

Sunnybrook and Women's College Health Sciences Center is the amalgamation of three of Canada's finest healthcare organizations. With over 1,400 beds serving acute- and extended-care patients, the healthcare provider has a history of excellence in patient care, education, and research. On an annual basis, the center experiences some 600,000 ambulatory visits, 51,000 emergency-room visits, and 28,500 inpatients; it also cares for approximately 3,700 newborns each year.

In terms of foodservice, the main complex has four retail outlets for employees and guests, plus one at the downtown facility. Patients enjoy excellent foodservice at all the locations. Of particular importance, the foodservice department is known for its high-volume catering activities. On any given day, the department caters 20 to 25 functions. These range from continental breakfast for the medical staff to elaborate weddings, most often held in the large conference areas that are part of the main complex. The department recently catered a multicourse barbeque dinner for 1,500 employees.

With the continued growth and popularity of the catering service, the foodservice managers have consistently battled with logistical issues. For example, coordinating rooms, dates, and times was a time-intensive exercise. And as with most on-site providers, the labor needed to perform such tasks was simply not available—particularly when labor was needed for catering the actual events.

The savvy foodservice operator, who has used various technological innovations in other parts of the department to increase efficiency, integrated EventMaster Plus, an event-management software package designed specifically to streamline tasks associated with catering. According to the managers, the package was selected because it is user-friendly, fast (in executing processes), efficient, and easy to use.

What does the software do? According to those most involved with Sunnybrook and Women's catering, it allows multiple users to access data simultaneously. Prior to integrating the package, it was not unusual for the department to double-book events accidentally. Inordinate amounts of time were spent rearranging things and trying to shuffle employees in order to adequately deliver food and related services.

The system also provides easy access to menu and pricing information as events are planned. This includes querying the database for staffing requirements and staff availability. The package also generates event reports pertaining to precosting, revenue forecasts, and sales trend information.

The operator initially integrated EventMaster Plus into the foodservice operation to organize the catering efforts. Since that time, the management team has noted a reduction in time and money previously needed to do what the system does automatically. The software has simplified all aspects of the back-office catering efforts. Of greatest importance, however, is that the system has resulted in greater satisfaction for customers—while saving the department money.

MENU PRICING

Pricing is addressed last in this chapter because it is a trickier issue than the other factors discussed earlier, especially when applied to catering menus. A set price, when the intention is to serve parties averaging dozens of guests, will be too high for customers looking for caterers of large events. Creativity and flexibility, mentioned earlier as desirable menu characteristics, also make pricing rather thorny. In addition, while customers are typically willing to spend more money when dining out, most will shop around for a caterer and make decisions based on a combination of price, affinity with the catering salesperson, and the reputation of the caterer.

In general, catering menu-pricing decisions are the result of two primary drivers, market and demand. Prices are market driven in that they must be responsive to the competition, particularly for menu items and delivery methods that are common across multiple providers. Demand-driven pricing can be adopted more fully for customers who want items or service styles for which there are few providers or alternatives in the marketplace.

Beyond these considerations, the traditional pricing approach is to use one of a handful of menu-pricing models.[2] The first of these is the nonstructured approach, known also as "seat-of-the-pants pricing." More quantitative approaches that are widely used include the factor method, the prime-cost method, the actual-cost method, the gross-profit method, and the stochastic-modeling approach.

Nonstructured pricing is the simplest approach to catering pricing. It involves a cursory examination of the competition's prices without regard for other factors. Needless to say, this method is completely inadequate. Surprisingly, some operators still use this approach.

The **factor method** is the most common one used in pricing and dates back more than 100 years to hotel-restaurant operators in Europe. A "factor" is established by dividing 1.0 by the desired food-cost percentage. (For example, a 37 percent food cost would result in a factor of 2.7.) The raw food cost for an item is then multiplied by the factor to produce the menu price. An item with a food cost of $2.25 would therefore be represented on the catering menu at a cost of $6.08. This is a popular method owing to its simplicity. (As many might note, an alternative calculation is simply to divide the raw food cost by the desired food-cost percentage.) The downside is that not every item should be marked up by the same proportion, since high-cost items would be priced beyond their market value. Also, low-cost items such as coffee, which normally produces a higher than usual **contribution margin** (sales less all variable costs or, put another way, the amount of sales that is contributed to fixed costs and profit), would be priced too low to generate an adequate return. Finally, this method does not reflect potentially differing labor costs associated with different items.

The **prime-cost method** is a variation of the factor method and integrates both raw product and labor costs. This method requires that labor costs be separated into direct labor costs (labor used for the preparation of a specific menu item) and indirect labor costs (labor used to finish the item, such as grilling, frying, etc.) for each menu item. The biggest advantage is that the prime-cost method reflects differences between labor-intensive items and prepared foods. The challenge is to make such allocations accurately, particularly since these may vary with the volume of items sold; this reflects the comment made earlier on pricing a catering menu that is versatile enough to meet the needs and expectations of customers who want both small and large functions.

Here is a simple example using only a single menu item. Assuming that the total direct and indirect labor costs for a steak sandwich (including the cost of benefits) is $2.60, the desired food cost is 37 percent, and the desired labor cost is 38 percent, the price of the item is calculated as ($2.25 + 2.60) / (37% + 38%) = $6.47.

The **actual-cost method** accomplishes the vital goal of including profit as part of every price on the menu. It uses food-cost dollars, total labor cost per guest, related variable cost percentage (covering such items as paper goods), fixed cost percentage (critical when space or equipment must be rented as part of the catering agreement), and the desired profit percentage. While limited to information found on the pro forma statement, this method is useful in that the inputs are more inclusive. The calculation, using fictitious numbers for a generic item, is:

	EP cost ($)	Selling price = X
+	Labor ($)	Var., fixed, & profit = 34%
+	Variable cost (%)	Food & labor = $4.85
+	Fixed cost (%)	Food & labor = 100% − 34% = .66
+	Profit (%)	.66X = $4.85
	Menu price	X = $4.85 / .66 = $7.35
		Therefore, the menu price is $7.35

The **gross-profit method** is intended to produce a specific amount of money that should be based on the number of guests attending a catered event. This is better than the previously discussed methods since the focus is on **gross profit** (sales less the cost of food); but because it is predicated on the accuracy of prior data and is specific to the number of guests, last-minute changes can affect the pricing structure dramatically.

Assume that the menu, offering sandwiches and one side dish, is intended for lunch parties of 100 and that the desired gross profit is 63 percent. Using data from other events, the caterer estimates that gatherings of this size generally produce revenue of $1,368 from food sales. Thus, the targeted gross profit is $862, or $8.62 per person. Assuming that the food cost for the specific menu is $4.25, the price for the menu is $12.87 per person.

This method underscores how important it is to price the menu on the basis of the number of guests. For this reason, some caterers offer tiered pricing for each type of menu. While this can be somewhat complicated and may limit versatility, it does make things clearer for customers.

The **stochastic-modeling approach** is the only method of pricing that integrates internal and external variables, particularly demand functions such as item popularity and market position of the caterer. In the event that off-site catering is a new venture for an on-site operator, the market position may be intentionally undervalued in the beginning. Conversely, if the reputation of the on-site staff is already recognized in the marketplace or if the products offered are unavailable elsewhere, the market position is considerable and should be reflected in the pricing algorithm.

The data necessary to price the menu using this approach are:

Oc = Percentage of sales allocated to costs other than food and labor (sometimes referred to, somewhat incorrectly, as "occupation costs")

Lc = Percentage of sales allocated to labor

Pm = Percentage of the menu price (MP) desired for profit markup (which may vary for different items)

EP = Raw food cost (also known as the "edible portion")

Mc = A subjective determination of market sensitivity to the catering firm; this is the variable representing promotional discounts or premiums

The menu price calculation, then, is

$$\text{Menu Price} = \frac{EP}{(100\% \ MP) - (Lc + Oc + [Pm \cdot Mc])}$$

As an example, assume that a customer wants a luncheon for 100 guests with an Italian-themed menu. The customer, an employee of the host organization, is looking at different caterers in town but wants to check with the on-site operator, who had just distributed a flyer about the department's new catering arm. Market analysis performed by the new caterer suggests that competing firms, particularly the one considered the best in town, promote menus appropriate for this event at a sizable markup. The on-site operator decides on a conservative profit of 10 percent but also wants to include a 5 percent discount (which means $Mc = 0.95$) from the total profit to promote the business. The resulting calculation, assuming 25 and 38 percent for Oc and Lc, respectively, and a raw food cost of $3.92, is

$Oc + Lc + (Pm \cdot Mc) = 72.5\%$;   therefore, the food cost is 27.5%
$MP = EP \ / \ (MP\% \text{ available for food}) = \$3.92 \ / \ 27.5\% = \$14.25$

Note: Thus, when Mc is less than 1.0, it indicates a discount; if it is greater than 1.0, the item should be sold at a premium, as in the case of a unique menu item.

Again, this list of menu-pricing approaches is not exhaustive. Some computer packages simplify the process, and some integrate a variety of other variables. The important thing is to understand how each of these variables functions in producing individual or whole-menu prices. And as with most management functions, menu pricing is both a science and an art; failure to embrace both will result in less than optimal operating results.

CHAPTER SUMMARY

In-house catering is most often offered to the host organization as a service rather than as a profit generator. Such a service builds awareness of the foodservice department's prowess and gives employees the opportunity to learn new foodservice-related skills. A number of problems must be solved, however, for an on-site foodservice operator to be successful in this area. First, facilities—particularly those in the back of the house—may not be sufficient to handle large catering jobs. An associated concern pertains to employees. Customers' expectations, built on what they have seen in the marketplace, are another hurdle.

Allocating internal charges—the method most commonly used to reimburse the foodservice department for its catering efforts—is the final major challenge for in-house catering. Even when costs pass through directly, it is essential that the operator accurately track and pass along all variable expenses for events.

Off-site catering offers many of the same benefits for the on-site operator but also includes the potential to substantially increase the bottom line. The first challenge most often faced in this area revolves around menu development. Sales and marketing is the next area that may be problematic.

The final major area pertaining to catering that is critical for on-site operators is menu pricing. Catering menu-pricing decisions are the result of two primary drivers, market and demand. In integrating these forces, pricing approaches range from overly simple to multivariate. Common approaches include nonstructured pricing, the factor method, the prime-cost method, the actual-cost method, the gross-profit method, and stochastic modeling, covering the gamut in terms of utility for pricing a catering menu.

KEY TERMS

environmental scanning
four Ps
nonstructured pricing
factor method

contribution margin
prime-cost method
actual-cost method
gross-profit method

gross profit
stochastic-modeling
 approach

REVIEW AND DISCUSSION QUESTIONS

1. For what reasons should an on-site operator offer in-house catering? What are the reasons against it?
2. What steps facilitate quality in-house catering in terms of human-resource management?
3. How important is environmental scanning? How often should it be performed?
4. What should be included when calculating the internal charge for a function when only variable costs are passed on to the in-house customer?
5. The importance of menu development for public eateries was discussed in Chapter 4. Is the level of importance different in developing a catering menu? What are the differences?
6. Give some examples of how the four Ps can be used to maximize an on-site operator's catering efforts.
7. How important is technology in project coordination? How might the number of catering activities impact the use of technology?
8. What is the difference between market and demand drivers in menu pricing?

9. Why would an operator choose to use nonstructured pricing?

10. What is the advantage of using the actual-cost method over the prime-cost method?

11. What is the advantage of including the Mc variable in the stochastic-modeling approach to menu pricing?

END NOTES

[1] Philadelphia: David McKay, Publisher, p. vii.

[2] A number of books and articles are available on this topic and do a good job of describing the many facets of pricing. One of the better sources is Miller, J. E., & Pavesic, D. V. (1996). *Menu pricing and strategy*. New York: Van Nostrand Reinhold.

INTERNAL CUSTOMERS AND SYSTEMS

Part III focuses exclusively on the end user, the external customer, in framing the discussion of key operational issues endemic to on-site foodservice management. This part assumes a different vantage point—one that seeks to analyze operations from the perspective of, or at least by giving consideration to, internal customers. Internal customers include employees, vendors, and supervisory staff. And while all tasks pertaining to an operation ultimately must be done with the external customer in mind, the role and responsibilities of internal customers require definition in order to make good on the ultimate delivery of food and related services.

To this end, Chapter 7 begins with back-of-the-house processes surrounding the handling of food. On the basis of accurate sales forecasts, purchasing deserves first consideration—this function involves vendors, a vital category of internal customers. Next, the nuances of ordering and receiving are discussed. This leads to the issue of inventory management.

Chapter 8 addresses the major processes affecting employees, arguably the most important category of internal customer. Recruiting, selection, training and development, and retention management are central to managing human resources appropriately. This discussion segues naturally into the primary topic of Chapter 9, productivity. Maximizing human capital receives considerable lip service in the hospitality industry, but quantifying and measuring productivity is a specialized skill, particularly in the complex on-site foodservice realm.

This part culminates with a dialogue in Chapter 10 regarding leadership and motivation. As any good operator knows, this area is critical to a manager's—and an operation's—success. Leadership and motivation pertain to all internal customers but concern employees more than anyone else. To offer the most robust information on the topic, the chapter includes both theory and the application of key principles in this area. The best practices in Chapter 10 further underscore the relevance and importance of leadership and motivation for on-site foodservice managers.

This part is critically important to managers in that it builds on much of the theoretical discussion offered in previous chapters. Of greater importance, it operationalizes many of the concepts discussed in foodservice textbooks from the perspective of on-site foodservice management. Finally, the best practices presented in this part—in locations as diverse as Overland Park, Kansas; Toronto, Canada; and Reus, Spain—highlight how managers are harnessing the potential of internal customers to maximize the delivery of foodservice.

CHAPTER 7

FOOD—PURCHASING, RECEIVING, AND INVENTORY MANAGEMENT

A leading objective of any foodservice operation is to handle food and beverage ingredients so that the final products both exceed customer expectations and are cost effective. Success in meeting these objectives, however, is predicated on successful implementation of several systems specific to the back of the house. Moreover, these systems entail different vendors and different members of the food-production staff.

This chapter offers a variety of perspectives on the best way to approach and integrate these systems with deference to the internal customers who are directly affected. Beginning with purchasing, the discussion focuses on accurately forecasting sales and then translating these forecasts into an identification of purchasing needs. An operation that uses many preprepared products is profiled as a best practice owing to the importance of forecasting and purchasing with the associated production system.

Issues surrounding receiving, including important topics such as food safety and coordination within the different components of the operation, follow. The chapter concludes with a discussion of the many facets of inventory management. A best practice featuring the application of a perpetual inventory system appropriately emphasizes the value of dedicating managerial resources to this vital function.

PURCHASING IN A CHANGING ECONOMY

At one time **purchasing,** defined as *obtaining the desired product that meets or exceeds specified quality standards for an on-site foodservice operation at the right time, in the correct amount, and at an acceptable price,* was relatively simple. Items were ordered to match a par level. When the quantity on hand fell below a certain level, commonly referred to as a **trigger,** the manager placed another order. It did not matter that the cost of carrying the excess inventory was prohibitive in terms of maximizing the operation's efficiency and profitability or that the likelihood of theft increased proportionately with the increased stock on hand. Moreover, the variety of products and the associated quality levels were relatively smaller, making price the key selection criterion.

With changes in the business world and a rapidly changing economy, such practices are viewed—appropriately—as unacceptable. Today, the purchasing process is complex and deserves considerable managerial attention. The key considerations applicable to on-site foodservice management include forecasting needs, ordering according to these needs (rather than par levels), and capitalizing on potential rebates.

Forecasting

As discussed in earlier chapters, almost every operational consideration begins with the menu. Properly planned, the menu reflects the available equipment on hand, the layout, and the target market. With regard to purchasing, the menu-planning process also reflects details regarding the necessary products.

Specifically, the part of the menu-planning process that speaks directly to the purchasing function is the formation of desired product specifications, which means identifying the type and quality level of a product. Should the menu item feature a 4-ounce or 5-ounce chicken breast? Should it arrive with or without bones? How about with or without skin? Is precooked, pulled chicken meat sufficient for the item?

This can be a lengthy process, depending on the number of products used in the overall menu and the availability of different types of each product. Other examples: Are fresh or dried herbs desired? What level of preproduction should be integrated into the product—should bread arrive freshly baked, or should it arrive as frozen dough that needs proofing and baking? Considerations pertaining to these questions again come back to unit-specific features (such as layout, equipment, and labor) and the price point of each menu item.

With the entire menu-planning and product-specification processes complete, the next step is **forecasting.** In its simplest application, forecasting means predicting the food items needed for a specific period, be it daypart, day, or week. Forecasting is important because it is vital to optimal financial management; it also precedes all other considerations regarding labor scheduling, space utilization, and, of course, purchasing.

Typically, forecasting sales—which translates directly into a forecast of what is needed—uses historical data such as the number of customers served per daypart, menu items sold per daypart, number of employees within the host organization, number of visitors to the host organization, and daily patient census (for healthcare only). Other information that is relevant includes changes in the competition, seasonality, increases or decreases in brand worth (for each brand represented in the operation), and changes in operational costs, such as dramatic swings in the price of specific food products, or, in terms of labor, changes in the labor market that might translate into lower retention (and increased labor costs).

Forecasting models are numerous and range from relatively simple approaches to methods involving compound calculations (typically performed on a computer). Determining which method to use depends on the desired level of accuracy, the availability of data, and the sophistication of the foodservice operator. A handful of the most commonly used methods, ranging from simple to complex, are simple averages, moving averages, weighted averages, modified moving averages, exponential smoothing, and causal forecasting.

Simple averages track trends over the length of time selected. This approach assumes that all the data are equally important, no matter what other operation-specific factors have changed. An example would be taking the sales of a chicken dish for every daypart during which it is offered and calculating the mean number sold per daypart. This can be done for each day of the week (e.g., using data from past Mondays to forecast sales for next Monday) or it can be calculated on a more aggregated basis.

The **moving average** approach uses specific time-series data. Groups of data, say sales of chicken cacciatore for every Wednesday lunch going back eight weeks, are averaged to produce a forecast for the following Wednesday (using the simple-average approach). The next forecast is calculated by adding the most recent number and dropping the oldest sales figure.

Weighted averages also use historical data but allow the operator to apply different weights to data points that are deemed more relevant to the current period. For example, suppose that a specialty fried shrimp platter, featuring large shrimp and offered every Friday, sold well in month 1. In month 2, however, a scarcity of large shrimp forced the operator to raise the price of the dish to offset the higher cost of the raw product; as a result, fewer orders were sold. In month 3, the operator lowered the price because there was an abundance of shrimp at a good price from numerous vendors; as a result, shrimp platter sales skyrocketed. In month 4, prices stabilized and the operator featured the item at the same price as in month 1. Sales leveled off, partially because customers disliked the increase in price over the previous period.

The operator then forecasts for month 5, during which he or she plans to lower prices slightly—somewhere between the prices featured in months 3 and 4—hoping that sales will be better than those in month 1. The weights, then, might be assigned as 20 percent for month 1 since this is a baseline for sales, 30 percent for month 2, 30 percent for month 3, and 20 percent for month 4. The resulting equation would be .20(month 1 shrimp platter sales) $\times$ .30(month 2 shrimp platter sales) $\times$ .30(month 3 shrimp platter sales) $\times$ .20(month 4 shrimp platter sales) = forecasted shrimp sales for month 5.

The **modified moving average** approach averages historical data, weighting the most recent data more heavily. For example, an operator might use the last four periods to project sales for this upcoming period but would add more weight to the two most recent periods. Such a weighting is particularly appropriate when attempting to compensate for such factors as seasonality.

Exponential smoothing uses a smoothing constant as well as recent actual data and the forecast for the past period to estimate future sales for an item. This is appealing because it allows the operator to gauge the accuracy achieved in forecasting the previous period and facilitates the integration of a correction factor into the equation. (This factor might also reflect promotion strategies.) The smoothing constant is a number between 0 and 1; it should be closer to 0 if sales have been relatively stable in the past and closer to 1 if the menu item is experiencing continued growth or if promotion for the item has been increased.

Using the shrimp dish example and operating under the principle supporting this approach that the forecasts will improve with each period, assume that the forecast for month 5 was 8,500 orders and the actual was 9,770. The general formula is:

New Forecast = past forecast + [smoothing constant
$\times$ (actual demand − past forecast)]

Assuming that the operator plans to promote the dish for the next period relatively highly, the forecast calculation might look like this:

$$\text{Month 6 Forecast} = 8,500 + .9(9,770 - 8,500)$$
$$= 9,643$$

Finally, **causal forecasting** approaches are the best and also the most complex. These approaches use multivariate statistical models integrating regression analysis to predict outcomes based on a number of input variables and also afford the operator the ability to control for certain variables. Specifically, causal forecasting allows an operator to integrate past sales data on each menu item to determine the forecast for a specific menu item, thereby accounting for the influence of one item on another. Since the presence of one menu item often influences the sales of another, this is advantageous. In addition, causal forecasting allows the operator to integrate advertising data—both past and present—as well as pricing strategies associated with the various menu items in the model.

Owing to its complexity, causal forecasting is not often used. In order to deal with the associated nuances, for example, dedicated statistical software is generally warranted. As foodservice managers become more computer savvy and general computer applications become more robust, however, this method will likely gain acceptance.

Ordering

Ordering is relatively straightforward, assuming that forecasts are accurate and production schedules are prepared accordingly. The delimiting factor is product delivery to the operation. If all items are delivered daily and in specific quantities, the process is simple. Usually, however, vendors deliver two or three times per week, and deviations from standard quantities carry extra charges. Unfortunately, this has led to a practice common in foodservice management known as the "just in case" approach; extra inventory is kept on hand just in case vendors are late in making deliveries, mistakes are made in the ordering process, items disappear from inventory, and so on.

Many large vendors have simplified ordering through automated systems allowing operators to place orders and receive information on product availability in real time. Thus, there is no longer a question of whether the shrimp ordered on Monday will arrive as planned on Wednesday. Some vendors even offer guarantees regarding the availability of certain items, making the foodservice manager's job easier.

The chief objective in ordering is to maintain a minimum amount of inventory on hand at all times. While more is said about this in the final section of this chapter, it is sufficient to say here that this is a generally accepted principle of quality foodservice operations. Exceptions include staples or regularly stocked items that benefit the operator when ordered in larger quantities or when the cost of placing an order is sufficient to persuade the operator to order less frequently.

The common approach to calculating optimal order quantities and reorder points is a process known as **economic order quantity** (EOQ), referred to as early as the 1930s as "minimum cost quantity." EOQ is essentially an accounting formula that determines the point at which the combination of order costs and inventory carrying costs are the least. Originating in the manufacturing sector, it is beneficial to almost every onsite operation.

Some operators may argue that a **just-in-time** (JIT) approach—a quality initiative designed to minimize unnecessary steps in production, labor, and costs achieved when all ingredients necessary for a menu item arrive just in time for the production process—is best for all situations. EOQ is actually an integral facet of JIT because it optimizes operations while providing logical ordering guidelines that are sensitive to different product types.

For example, assume that an on-site foodservice operation forecasts 1,100 orders of veal parmigiana for a particular meal. According to historical data, approximately 300 orders are usually purchased to go and are therefore packaged in disposable containers. It is intuitively cost effective to have the veal arrive as close to the preparation time as possible. It is not cost effective, however, to have the exact number of disposable containers used to package the dish delivered along the same timeline; these are usually purchased in bulk and are used for a large number of different items. To determine the most cost-effective quantity of disposable containers (and the associated order time) the following EOQ formula is ideal:

$$EOQ = \sqrt{\frac{2UFx}{Hc}}$$

where

U = Annual usage
Fx = Fixed cost associated with placing an order
Hc = Annual holding cost per unit

Using the example of disposable containers, assume that the operation uses 275 cases of disposable containers per year and that the cost of placing an order with the vendor is $22 (including the labor involved in ordering and receiving). Furthermore, assume that the cost of carrying inventory, which entails insurance, storage costs (particularly if the item needs refrigeration), and typical loss due to damage, runs around $25. The EOQ calculation is

$$EOQ = \sqrt{\frac{(2)(275)(22)}{25}} = 22$$

Thus, the optimal ordering amount is 22 cases.

It is important to note that EOQ considers only the holding and order costs. It is based on even (and therefore known) demand, and the cost of goods is fixed at any order level. Finally, there is an explicit trade-off between the cost of carrying excess inventory and the cost of placing additional orders. While there are challenges associated with these assumptions, the value of EOQ is considerable. The key is to routinely recalculate the EOQ and to reevaluate ordering quantities accordingly.

Rebates

The idea of saving money by purchasing larger quantities is commonly understood in foodservice circles. It's a matter of economics. Many operators, particularly those working in noncontracted environments, have made alliances with others to maximize their purchasing clout and qualify for even greater discounts. In fact, this practice has resulted in the emergence of a variety of **group purchasing organizations** that facilitate the sharing gained from the resulting economies of scale.

Consider next the often overlooked but closely related topic of **rebates.** Provided by vendors, rebates are a reward system for operators who exceed certain purchasing thresholds. While the unit price may be the same for 100 cases as it is for 500 cases, some vendors offer a rebate when the operator purchases the 500 cases. Thus, incremental discounts are not given, but rebates that often equate to a larger dollar amount are provided.

Rebates benefit both vendors and operators. For the vendor, a rebate program offers an incentive for an operator or group of operators to buy a certain volume of products. In turn, the vendor can negotiate lower prices with food manufacturers. If the operator fails to meet the threshold for the rebate, the vendor is protected from loss resulting from having to carry the excess inventory. For operators, such programs allow for greater overall savings. The key is to forecast accordingly and to maintain consistent relationships with key vendors.

Some argue that rebates are nothing more than kickbacks. While this might have been true in the past, today's economy requires unprecedented managerial creativity. Rebate programs reflect this and offer new ways for on-site foodservice managers to reduce food costs in somewhat nontraditional ways. Rebate programs are not instituted for selected clients only; they are available from almost every vendor and can be used by any operator who is savvy in the various back-of-the-house management techniques.

RECEIVING AND STORAGE

The practice of receiving and storage is fundamental to the success of any on-site foodservice operation. We highlight here key aspects under the following headings: essentials of receiving, critical storage issues, and the key challenge of effective receiving and storage.

Essentials of Receiving

While reliable employees are important to every facet of any foodservice operation, staff members with specific skills or training are essential to some tasks, particularly receiving. In smaller operations, managers often prefer to perform this task themselves—when it is feasible—owing to its importance.

The employee responsible for receiving goods must know something about food. This may seem obvious, but too often receiving is delegated to employees who otherwise are charged only with sanitation. The person must also have the maturity to question things that appear incorrect or improper. Too many employees will take the word of a delivery person over their own observations. This is not to imply that delivery people are dishonest; rather, they are usually under considerable pressure to expedite deliveries as quickly as possible. Thus, they may not be as attentive as they would be if they were not rushing.

From the operator's perspective, the onus is on the receiving person to ensure that what appears on the invoice is what is received, that the invoiced items are the same as those ordered, and that the goods are not damaged. To that end, the final attribute that receiving personnel must possess is integrity. As discussed in more detail later in this chapter, the receiving function exposes the operator to considerable risk. An employee who isn't trusted should not be charged with the responsibility for receiving. This is not to say that managers should not supervise the overall receiving process. The truth is that even the best manager can't scrutinize every aspect of the operations all the time. This is why employees' integrity is key.

This brings up the interesting practice of blind receiving. One version of blind receiving—advocated in some textbooks as a positive internal control process—is not to tell the receiving person what was ordered. In this way, the employee has only the invoice as a reference. This is a dangerous approach because it is wholly predicated on the accuracy of the invoice. In other words, an employee may verify that 50 pounds of beef tenderloin were delivered but will be unaware that it was pork tenderloin that the chef ordered. The other approach to blind ordering is to request from vendors invoices that have no weight or quantity information, thereby forcing the receiver to calculate the information. This arguably has some value, but it creates other issues that may arise with the interchange between the delivery person and the receiver. In most settings, the benefits of this type of blind receiving rarely outweigh the risks.

The one piece of equipment that is most integral to the receiving process is an accurate scale. Large on-site operations that take the receiving process seriously use scales capable of printing out each reading on a label that is then attached to each case or bundle. This process emphasizes to delivery personnel that others will also verify weights. Of course, the utility of a scale is assured only if it is used.

The final essentials of receiving are specifications and sanitation. If specifications are not maintained, the receiver has no point of reference. This is unfair to the employee and weakens the implicit message that receiving is a critical function. Moreover, specifications help educate the receiving personnel about what the operation orders and expects.

Equally important is sanitation. Receiving areas should be cleaned regularly and treated as an extension of the kitchen. The receiving dock should not double as a smoking area or a storage area. It is a functional space that should reflect its level of importance in the way it is maintained.

Critical Storage Issues

Most issues associated with storage are fairly straightforward. Receiving areas should be large enough to accommodate full orders. There must be sufficient space for "dry"-goods storage as well as refrigerator and freezer space. The temperature and ventilation of the dry-goods storage areas should be adequate. Finally, all storage areas should be clean and orderly.

Critical issues that are too often overlooked include labeling and rotating stock. Labeling means dating everything when it arrives, regardless of whether an item is refrigerated, frozen, or canned. Even cases of soda should be labeled with the arrival date.

Dating achieves several goals. First, it emphasizes to employees that management is serious about effective food safety and handling. Second, it underscores how long items or ingredients sit on shelves. After all, it is easy to forget that an item has been in the freezer for an extended period of time. When an operator sees the date every time he or she enters the walk-in storage area, however, it is hard to ignore.

The third goal of dating all inventory items is tied to the other critical storage issue, stock rotation. Most operators are familiar with the **first in/first out** (FIFO) method of inventory rotation, in which items are used in the order in which they are received. The problem is that the success of a FIFO program rests entirely on effective dating. For example, canned goods received yesterday, assuming they are the same brand and the logo or packaging has not changed, look just like the ones received three months earlier. The only way to differentiate them and to ensure that FIFO is used successfully is to date everything.

An application of FIFO's critical importance occurs when numerous prepared products, some with a limited shelf life, are integrated (as discussed at length in Chapter 3). Such an approach reduces the total number of products to be stored since many are already combined; yet this necessitates thoughtful ordering and storing. One operation that has mastered this is Menorah Medical Center.

BEST PRACTICES IN ACTION

PRODUCT MANAGEMENT AT MENORAH MEDICAL CENTER (OVERLAND PARK, KANSAS)

The Menorah Medical Center, part of the Health Midwest family of hospitals, has a reputation for integrating cutting-edge approaches in healthcare. The Menorah campus includes a 158-bed acute-care hospital with the latest in medical equipment; it also boasts a 110,000-square-foot medical office building featuring many of the Kansas City area's finest primary-care, specialty, and subspecialty physicians and was designed for optimal outpatient care. Many have described the Menorah as the hospital of the future, largely owing to its design and its approach to healthcare.

Innovation can also be found in the foodservice department. The management team recently adopted a "speed-scratch" approach using many prepared items, some of which are service-ready and others that are components of various menu items. For ex-

ample, low-sodium mashed potatoes are purchased in prepared form, which arrive in multiserving bags. In its original state, the potato product consists of freshly mashed potatoes and has a shelf life of seven days. As shown in Figure 7.1, employees need only transfer the product to a pot for warming and then can add selected seasonings to the base product to customize it. Cooks may add garlic and horseradish for introduction as a side dish in the cafeteria or rosemary, pepper, and salt as a specialty item for patients.

Using forecasts, products are ordered and production schedules are generated accordingly. If a forecast is incorrect and not enough product is available, cooks require very little time to prepare more of the high-demand item. A wonderful example is beef stew. The beef and sauce arrive prepackaged and precooked. To make the stew, a production employee can add prepackaged frozen carrots, prepackaged and precut potatoes, and the appropriate spices (specified in the standardized recipes). Including heating time, the entrée can be made in less than 15 minutes. Similarly, the biscuits that accompany the stew arrive as a "scoop and bake" product that requires little baking time.

Simplicity in ordering and preparation comes with a higher food cost, but the entire food-production process is greatly simplified, resulting in labor savings and economies associated with the production process. In addition, the end products are much more standardized than menu items that are subject to more complex methods of preparation. Finally, the requisite skill level of the production staff need not be very high, given the simple pro-

duction methods. Menorah has developed the food processes so well that employees know what products should look like when they arrive in the receiving area and, through colored pictures posted in the production area, know what the dish should look like when it leaves the kitchen.

The savings from this approach, particularly given the manner in which the management team has executed it, are obvious. Less obvious savings stem from efficiencies in purchasing, receiving, and storage. Fewer products come in the back door, making each of the aforementioned steps much easier. Moreover, management is able to keep a firm grip on what leaves the kitchen, and is better able to ensure that the food that does leave will produce high levels of customer satisfaction.

FIGURE 7.1 Preparing Mashed Potatoes at Menorah Medical Center. Copyright by Sodexho USA.

The Key Challenge of Effective Receiving and Storage

The key challenge associated with the receiving and storage process is to ensure that the flow of goods is as efficient as possible. This involves minimizing the opportunity for employee theft, often referred to colloquially as **shrinkage.**

The reasons for employee theft are too numerous to discuss in a single textbook. It is sufficient to say that management must minimize the opportunity for theft. This principle has many permutations, including *keeping honest people honest.* Another way to look at the objective of minimizing the opportunity for theft is an old Ronald Reagan adage: *trust but verify.*

The first opportunity for theft is potential collusion between the receiver and the delivery person. This is often overlooked since some think it an unlikely marriage. Nonetheless, losses due to this type of collusion can be huge. The key is to routinely spot-check invoices, examining orders in part or in whole. The implied message should be that management is always looking to ensure that things are going well, not that employees aren't trusted.

The second opportunity for theft is in the transition from the receiving area to the storage area. Items that are not placed in the correct storage area are the ones most susceptible to theft. Remember that food is an asset that has broad-based appeal; everyone uses it, and everyone knows the value of expensive items.

The final area is the effective management of the inventory. The notion that a person uses more toothpaste when the tube is full has direct application to the food-storage process. When employees see large quantities of a certain product, they tend to attribute less importance to it. Thus, an unnecessarily high volume of stock on hand translates into unnecessary costs and unwarranted exposure to potential shrinkage. This issue of effective inventory management is discussed further in the next section.

INVENTORY MANAGEMENT

Inventory management is critical to managing the large volume of different menu items in most on-site foodservice operations. The best approach is to maintain accurate inventory records, to ensure that control measures are in place, and to integrate sound inventory-turnover analysis.

Inventory Records

Inventory record keeping begins with the receiving process. Following product inspection and invoice-accuracy verification, the relevant data must be compared with ordering information to ensure that the prices are correct. Next, the data must be updated in the inventory log.

This inventory log is an integral part of a **perpetual inventory** system. Critical to optimal inventory management, a perpetual inventory is a running record of what products are on hand; the inventory is updated in real time through integration with point-of-sale terminals and requisition orders, as well as with information from invoices on what was received. In the event that minimum and maximum quantities of each inventory item are established, a complete perpetual inventory system will flag items that need ordering or that are overstocked.

Perpetual inventory systems were once considered too labor-intensive for on-site operations. That perception, thankfully, is changing. Through sophisticated software packages, the implementation and maintenance of perpetual inventories offers myriad advantages. A best practice that underscores the value of using this approach is featured below.

FLORIDA HOSPITAL'S PERPETUAL INVENTORY (ORLANDO, FLORIDA)

The foodservice department in the Florida Hospital in Orlando, which is part of an expansive healthcare system serving central Florida and the surrounding counties, is recognized for its efficient record keeping. As in many other large hospitals, this is simply good business. What makes Florida Hospital in Orlando unique is that in addition to serving food to patients in its 881-bed acute-care center and 53,600 outpatients who visit the center each year, it prepares food—everything from shredded lettuce to soups to

FIGURE 7.2 A Staging Area at Florida Hospital in Orlando, Florida. Copyright by Florida Hospital.

meat loaf—for seven other hospitals in the system. In total, the eight hospitals include some 1,750 beds and service approximately 275,000 outpatients annually. The enormous volume in the different food outlets matches these impressive numbers; one of the staging areas is shown in Figure 7.2.

According to the foodservice management team, the key to maintaining effective inventory management is their elaborate perpetual inventory method. Here's an example of how it works: The foodservice manager at the Winter Park Memorial Hospital forecasts that they will need 20 bags of split pea soup (each bag weighing 12 pounds) on Monday of the following week. They complete the electronic requisition form, which is automatically transmitted to the foodservice department in the Orlando location. The computer system records the order and schedules delivery of the soup. When the delivery is prepared, employees record what they have taken from the prepared stock. The system uses the data to issue production reports ensuring that more split pea soup will be produced in anticipation of forecasted demand for the menu item (which includes all eight hospitals). What is more impressive is that the perpetual inventory system automatically generates order sheets so that the necessary raw ingredients can be ordered according to a timeline that ensures that they are on hand according to the production schedule.

Given the huge number of products that are generated by the central facility, including many items prepared using cook-chill equipment, keeping a handle on inventory management would be a challenge for any operator. Using technology and good management practices, however, the team has refined the perpetual inventory system so effectively that no one can imagine doing it differently; it just would not be as efficient.

The hospitals that use the food are also in the process of developing their own perpetual inventory management methods that will eventually be tied to the centralized kitchen, further streamlining operations systemwide. And given the likelihood that the healthcare system will continue to grow, and the economies of scale that have been realized through the centralized approach, the perpetual inventory system may be the most important management tool in their business.

Even with a perpetual inventory system in place, a physical count of items in all areas of an operation is critical to maintaining accurate records. A **physical inventory,** which in on-site foodservice is usually taken weekly, is simplified by the many inventory management software packages available. (A solid perpetual inventory system, the primary purpose of which is to verify amounts on hand, can make a monthly physical inventory sufficient.) Even the simplest of these, usually nested within spreadsheet packages, integrate a physical inventory form that can be used to record the information manually. The data are then entered into the computer program. More sophisticated programs involve entering the information in wireless handheld computers that automatically update records.

The physical inventory process is critical to maintaining accurate records. It is also an aspect of the managerial process that is fraught with opportunities for mistakes and, potentially, the chance for dishonest employees to hide past or planned thefts. Thus, inventory records must be tied closely to inventory control measures and diligent inventory-turnover analyses.

Control Measures

Inventory control measures are critical to the financial success of any foodservice operation.[1] First, maintaining the smallest investment in inventory is prudent. Inventory is a current asset that provides no return on investment until is it prepared and sold.

Thus, excess inventory is a nonperforming asset. Moreover, fewer items in stock equate to less work in storing and protecting the asset. This is underscored by one finding that suggests that every dollar of inventory represents at least $25 per year in terms of expenses related to financing, handling, storage, and insurance.[2]

Excess inventory can affect profits negatively in other ways. As noted earlier, one of the difficulties in dealing with food is its high portability. Another problem is its universal utility. (In other words, too much inventory provides unnecessary opportunities for theft.) Furthermore, excess inventory leads to increased labor costs owing to the need to rotate and handle the food. Inflated soft costs such as utility expenses associated with refrigerators and freezers can also result from overstocked inventories.

The trade-off for inventory management is having enough food on hand so that menu items are always available. This supports the old adage from manufacturing that says that inventory management is all about carrying stock versus the risk of **stockouts.** In foodservice, stockouts mean that customers' expectations are not met and the entire dining experience is diminished. Equally bad, inappropriate recipe substitutions are sometimes made when ingredients are not in stock, resulting in low-quality products that diminish customers' confidence in product consistency.

The second control issue pertains to providing secure and quality protection for all food products. Returning to the earlier discussion of storage, a key to control is limiting access to the various storage areas to those who need access. These areas must also be maintained at optimum levels of cleanliness, temperature, and humidity.

Next, accurately valuating inventory is key to accurate control. Given that the food cost is determined by adding purchases to the value of the opening inventory less the value of the closing inventory, valuation is at the core of all accounting functions related to food. This means that prices must always be updated. It does not matter that a can of stewed tomatoes was purchased for $5 six months ago; what is important is the cost of replacing the item—its value today.

The other part of accurate valuation is accurate counting, part of the physical inventory process discussed earlier. Controls in this area require that (1) at least two people perform the physical inventory and that (2) the responsibility for this task is rotated among appropriate staff members. Such safeguards minimize the likelihood of collusion while reducing the chance for a single employee to use the inventory-taking process to hide pilferage.

A good control tool that helps ensure that potential problems in the physical-inventory process are minimized is the **inventory audit.** An inventory audit is a recount of randomly selected items immediately following the physical inventory process. Members of the management team usually perform the audits.

Random audits make sense because of the "80-20 rule." It is often said that 80 percent of the inventory cost—including the cost resulting from pilferage—involves only 20 percent of the inventory items. (More is said about the origin and application of this rule in Chapter 11.) Thus, audits are typically conducted on more expensive and universally popular items such as meats and alcoholic beverages. Inventory audits can also be conducted at different times between the scheduled physical inventory dates in cases where specific items have been subject to mysterious shrinkage.

Inventory-Turnover Analysis

While the aforementioned controls certainly are requisite aspects of inventory management, one of the best tools used to monitor the overall process is inventory-turnover analysis. Inventory turnover can be calculated as follows:

$$\frac{\text{Food cost for the period}}{\text{Average inventory value for the period}} = \text{Inventory turnover for the period}$$

For example, assume that a modest-sized on-site operation has weekly retail sales of $47,000 with an associated food cost of $16,450. For the targeted week, the average physical inventory was $17,312 (i.e., the inventory's beginning and ending valuations divided by 2). The inventory-turnover statistic for the period is

$$\frac{\$16,450}{\$17,312} = 0.95 \text{ turn per week}$$

Of course, a single week's turnover statistic is not a valid benchmark for assessing overall inventory-management success. A more useful measure would be to calculate the inventory-turnover statistic over a multiweek period. Such a calculation is easily applied to categories of food and beverages and even to individual food items in cases where problems with the inventory-management process are severe (in much the same way as the inventory audit is used). Such analyses are especially helpful in identifying unexplained spikes in item-specific use, particularly for high-priced items.

The utility of the inventory-turnover statistic is perhaps most obvious when comparing a single operation over time with a number of very similar operations. Even if an operator regularly achieves similar levels of turnover, anomalies may indicate cause for concern. In Table 7.1, for example, the average monthly inventory-turnover statistic for the hypothetical medical center is within the target range of from 2.5 to 4 for healthcare foodservice operations. (This range is also applicable to education and corrections; in B&I, the range is 3.5 to 4.5.)

The data shown in Table 7.1 provide a good example of the turnover statistic's value, particularly when monitored over time.[3] For the first six months of the period under scrutiny, the operator maintained relatively consistent inventory turnover. For the next several months, however, a disturbing pattern emerged. The inventory turnover for months 7 and 8 was only 3.5 turns per month and declined during the next four months.

This pattern of inconsistency signifies erratic inventory management, which is undesirable under any circumstances. Moreover, the decline beginning in month 7 may indicate an unexpected change in sales or may reflect other operational problems; in either case, the information suggests that immediate investigation is warranted. Finally, the decline in turnover may be the result of employee theft or collusion in the inventory-valuation process. Regardless of the cause of the decline, the periodic analysis indicates that management should investigate other operating statistics to identify and remedy the problem. Granted, the cause may be innocuous, such as a change in vendor accompanied by a new delivery pattern. The point is to know when something out of the ordinary is happening and to be prepared to do something about it.

TABLE 7.1

Inventory Turnover Statistics for ABC Hospital

Month	Turns per Month
1	3.8
2	3.9
3	3.7
4	3.9
5	3.8
6	3.6
7	3.5
8	3.5
9	3.3
10	3.4
11	3.5
12	3.1

The data in Figure 7.3 indicate the occurrence of a cyclical phenomenon in relation to the inventory-turnover declines. That is, every three months, the inventory-turnover statistic takes a notable downturn. This can be explained if the same employee (or pair of employees) had a hand in the inventory process during those periods and is either making consistent errors or, more likely, is attempting to hide fraudulent activity.

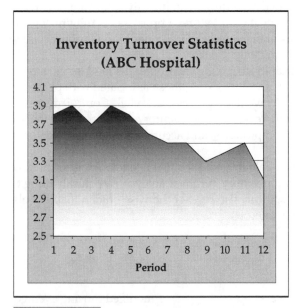

FIGURE 7.3 Graphical Analysis of ABC Hospital's Turnover Statistics.

Owing to the related managerial adage and perennial objective of "slow and old or fresh and fast," inventory-turnover analysis can aid operators in identifying problems related to inventory management and can directly help reduce associated costs. Such tools can also be critical in identifying situations normally thought of as highly unlikely (such as a favored employee stealing food). Without question, the growing complexity of today's on-site operations calls for innovative tools and approaches. This one deserves to be in every manager's toolkit.

CHAPTER SUMMARY

Purchasing has evolved over time from a simple procedure to a complex process. At the core of this process is the forecasting of sales, which then dictates what items are needed and when. Forecasting models include simple averages, moving averages, weighted averages, modified moving averages, exponential smoothing, and causal forecasting. Which to choose depends on the importance of accuracy, the availability of data, and the sophistication of the foodservice operator.

Optimal ordering, particularly for staples, is best achieved through EOQ, essentially an accounting formula that determines the point at which the combination of ordering costs and inventory carrying costs are the least. Rebates are also important for realizing maximum savings through quantity ordering.

Receiving is another key factor affecting several other processes. Receiving should be performed by employees who are knowledgeable about food, who possess the maturity to question things that appear incorrect or improper, and who demonstrate integrity in addition to the other characteristics of valued employees. Once received, food items must be labeled and rotated according to an inclusive FIFO approach. The main challenge in managing the receiving and storage processes is minimizing the opportunity for theft. This includes anticipating the potential for collusion between foodservice employees and delivery persons.

Inventory management is critical in managing the numerous food items used in most on-site operations. Inventory records, including those associated with perpetual inventory systems and the physical inventory process, are vital. Control measures such as keeping a minimal but adequate amount of inventory on hand, providing secure and quality protection for all food products, and accurately valuating inventories are similarly critical.

Inventory-turnover analysis is a valuable tool for managers looking to monitor the inventory-management process. Inventory-turnover analysis can quickly bring to the fore inefficiencies in the process. It can also reveal otherwise unidentifiable problems before they become financially onerous.

KEY TERMS

purchasing	simple averages	modified moving average
trigger	moving average	exponential smoothing
forecasting	weighted averages	causal forecasting

economic order quantity   rebates   physical inventory
just-in-time   first in/first out (FIFO)   stockouts
group purchasing   shrinkage   inventory audit
   organizations   perpetual inventory

REVIEW AND DISCUSSION QUESTIONS

1. Why is the practice of ordering everything according to par levels no longer used?
2. For what products should an operator develop specifications?
3. Of the many forecasting methods presented, which one holds the promise of greatest accuracy? Would you use it?
4. Assume that a foodservice department uses specialty disposable cups in the espresso bar, which results in approximately 110 cases of cups used per year. Because of the specialty logo, each order costs $50. Storage costs average $22 per year. Using the EOQ approach, what is the optimal order amount? What does this mean in terms of how many times to order per year?
5. Are rebates a viable approach for operators to lower their food cost? Explain your answer fully.
6. What are the key elements in receiving?
7. Why is the FIFO method important for foodservice operations?
8. What are the advantages of using a perpetual inventory system? Why wouldn't you use such a system?
9. Why are stockouts a problem for on-site operators?
10. Is it possible to turn over inventory too often? Provide examples with your answer.

END NOTES

[1] A variety of effective internal-control measures are discussed well in Geller, A. N. (1991). *Internal control: A fraud prevention handbook for hotel and restaurant managers.* Ithaca, NY: School of Hotel Administration, Cornell University.

[2] Wiersema, W. (1998). Inventory: How much is too much? *Electrical Apparatus, 51*(7), 40–42.

[3] This example is adapted from Reynolds, D. (1999). Inventory-turnover analysis: Its importance for on-site food service. *Cornell Hotel and Restaurant Administration Quarterly, 40*(2), 54–59. The article also provides other information about inventory turnover analysis that is beyond the scope of this discussion.

CHAPTER 8

HUMAN-RESOURCE MANAGEMENT

Of all the issues that face foodservice management, labor is undoubtedly the most complex and problematic. Pick up any trade journal and such issues as labor shortages, escalating labor costs, and other related topics will assuredly be included. This chapter is intended to provide some remedies, often case-specific, to many of the human-resource issues that plague most foodservice operations. Healthcare provides an excellent milieu for such a discussion, given the almost blinkered focus on costs that healthcare has witnessed during the past several years.

The first of these labor-related issues is planning. With its many facets, the planning function is critical, as demonstrated in the associated best practice. Following a strategic planning approach, the human-resource management process—and our discussion—logically proceed to the matter of job analysis. Next, the fundamental tasks of recruiting and selecting the best employees from the available pool of candidates are addressed. And while selection is critical, so is the training and development of the most valuable asset to any on-site foodservice environment: its people. The importance of this topic is underscored by an associated best practice. The chapter concludes with a dialogue about retention management, which includes issues surrounding the calculation of turnover.

PLANNING

It is no surprise that the foodservice industry faces a greater challenge than others in filling and managing hourly positions. How does a foodservice operator attract and retain reliable employees when the jobs feature long and irregular hours, lack of advancement, constant pressure to perform, and disruptions of family life from working holidays and weekends? Even in segments such as B&I, where the hours are more regular and typically do not include weekends, it is tough to find sufficient numbers of suitable employees.

For this reason, the **planning process** is integral to ensuring that an operation has a full staff, that changes in staffing levels are accurately anticipated, and that labor costs are not unnecessarily inflated by such unacceptable practices as overtime pay. Planning, in its simplest human-resource context, is the process of forecasting human-resource needs and determining the activities required to meet those needs. The driving purpose underlying the process is to link the human-resources element with the other aspects of the operation—from a strategic and operation-based context—and ultimately to strengthen the bottom line by enhancing the effectiveness and efficiency of all human-resource-related activities. To this end, there are four primary components of the planning process: environmental scanning, assessing human-resource demand and supply, designing and implementing associated programs, and evaluating outcomes.

Environmental Scanning

Applied externally in Chapter 6, **environmental scanning** used internally is the process of studying the organization's setting, situation, and business atmosphere to pinpoint opportunities and threats as they apply to the operation's human resources. This is often done using a SWOT analysis, in which the strengths, weaknesses, opportunities, and threats are quantified and analyzed both individually and with respect to the other issues facing the operation.

This part of the planning process is pivotal in that it sets the stage for future action. If performed correctly and objectively, environmental scanning can provide a wealth of information. If done incorrectly, or by managers who are unwilling or unable to identify their operation's weaknesses, the process will yield meaningless results and can be more harmful than helpful since the poor information may be used to make poor decisions.

Assessing Demand and Supply

Assessing demand accurately depends on accurate forecasting. This goes beyond forecasting sales and then matching staffing levels accordingly. It also means forecasting the skills needed for specific activities. For example, if a strategic objective for the upcoming period is to increase catering sales, then the question becomes one of determining what is needed to accomplish this. Does the operator need someone with better sales skills, or does he or she require a more skilled production crew?

Demand is usually assessed through one of two techniques. The first, the **expert-estimate technique,** is a bottom-up forecasting approach in which a manager estimates the needs of the operation based on past experience. This technique can also be applied to a group, in which select members of the management team contribute to the decision process. In some instances, this can be very accurate. The disadvantage is that it is solely dependent on the relative experience of the managers involved in the process and the prowess each possesses regarding staffing. Not surprisingly, the result of this approach is usually less than optimal.

The second is the **trend-projection technique.** This begins with the identification of key business factors that relate to staffing needs (e.g., sales). Management then calculates the relative productivity of each position based on key factors such as typical skill levels and abilities. The results of these calculations are often charted for comparison purposes. Next, historical data are used to project future demand based on key indicators such as sales (see Chapter 7 for the particulars of forecasting sales). Finally, the projections are adjusted for unusual events that have unduly impacted the operation or are expected to in the future.

Assessing the labor supply involves identifying internal and external opportunities. For the internal supply, managers must identify who is ready for promotion, who needs to be reassigned, and who is leaving. Information gleaned from supervisors as well as from performance appraisals is very useful in answering such questions. A related approach is to formulate a skills inventory and career aspirations for the entire staff. Such information greatly aids in making inferences about the internal supply.

The external supply is best evaluated by using external environmental scanning. While the economy and the associated unemployment rate are often the biggest factors in determining labor availability, factors affecting certain sectors also must be considered. Furthermore, evaluations of immediate competitors, both within the on-site segment and in the hospitality industry at large, can help in assessing the external supply.[1]

Designing and Implementing Programs

With needs assessment and knowledge of the supply in hand, managers next must design and implement programs that will achieve the business strategy's objectives. Where there is a shortage of human resources, for example, programs directed toward recruiting and quality selection are key. Similarly, the right combination of needs may dictate that more resources should be dedicated to training and promotion. In extreme cases—say, in the event that an adequate number of cooks, or those of appropriate caliber, cannot be found—the operation may need to rethink its many processes and possibly introduce more preprepared products.

In cases where there is a surplus of human resources, usually leading to downsizing within the host organization, the on-site foodservice manager has a number of programs from which to choose. These might include reducing the number of staff members through attrition. Early retirement programs are often useful when integrated in tandem with such approaches. When the need to downsize is more immediate, management may demote substandard performers. A more drastic approach is to terminate those individuals who fill positions that become unnecessary during sales downturns.

A best practice that stands out, in large part due to its unique situational factors, underscores how the total planning process can be implemented despite a variety of monumental challenges.

BEST PRACTICES IN ACTION

HOSPITAL UNIVERSITARI DE SAN JOAN DE REUS (REUS, SPAIN)—THE PRACTICE OF SELLING, NOT TELLING

While healthcare in the United States has seen and continues to see its fair share of fiscal woes, the problems funding medical care in Spain dwarf those in almost every other country. Healthcare in Spain faces enormous labor costs. Union employment prevails, with most unions possessing incredible control and power over labor agreements. This is particularly problematic for foodservice, which is largely labor dependent.

Simply stated, it is difficult to introduce changes that ultimately reduce labor needs. For example, a typical foodservice employee who is dismissed for almost any reason—even in the case of forced downsizing—may receive up to 45 days of pay for each year worked upon termination; this amount is calculated using the rate of pay received at the time of separation from the employer. This means that an employee who has worked for 10 years and who makes $10 per hour would receive $36,000 if terminated. Consequently, causes for dismissal are very limited and must be supported with ample documentation. Needless to say, simply managing the foodservice labor force can be daunting, particularly for less experienced managers. Combine this with a business environment that is particularly short-term in focus, at least in terms of fiscal management, and the problem is obvious.

The challenges don't stop there. Given the high cost associated with terminating employees, the tenure of most hourly workers, especially in support services, tends to be very long. The collateral issue is that employees often assume a less than optimal attitude, given their high degree of job security. As a result, any change is difficult to implement, particularly if it results in increasing the workload.

In sum, this means that the process of planning must be done almost in reverse. Since changes in staffing are so problematic, management must start with this reality and then consider what operational strategies align accordingly. This also raises issues related to launching new initiatives, since the labor force is so deeply entrenched.

The foodservice managers at the Hospital Universitari de San Joan de Reus, in Reus, Spain, broke the mold in many ways by breaking away from the traditional management style of "telling" employees what to do regarding the strategic plan and the associated human-resource planning process. Instead—and their reasons related largely to their goal of alleviating issues that would force them to eliminate positions and incur short-term hits to operating income—they did what many consultants do and asked the employees for input on various issues regarding the operation. For example, patient satisfaction related to food quality had been a hot topic for some time. Thus, they asked employees for their opinions, both about the problem and about potential solutions.

With ideas for remedying some problems, the foodservice managers, led by the creative foodservice director, developed a workable strategy and then went about the task of "selling" changes that would positively affect outcomes related to the identified food issues. Is this rocket science? No. But the outcome was positive; in some cases, it even resulted in employees assuming more responsibility or changing their routines, which often involved more work on their part.

What makes this solution so impressive are the situational factors in the foodservice department. The employees have long maintained the attitude that "We've seen them (managers) come and go— yet we remain, doing what we know best." This approach, then, represents a big first step in changing the status quo, and doing it with buy-in from all constituents. The primary outcome, of course, is that it gets everyone thinking about the customer— something that is not always easy to do.

Evaluating Outcomes

The key to good planning is continuous evaluation of each phase of the planning process. For example, assessing the demand and supply forecasts is accomplished through periodic checks of actual versus predicted staffing levels. Another useful analytic tool is the tracking of implemented programs versus predicted program requirements. In other words, did the foodservice department do what it planned? This approach is particularly useful in evaluating the implementation and effectiveness of downsizing through attrition.

Another means of evaluating the outcome of the planning process is to compare human-resource programs and the associated labor costs with budgetary figures. It is easy to overspend on some initiatives. Constant vigilance is vital to keeping spending in check. Finally, the most important evaluation to make is the comparison of performance and service levels with predicted goals. If staff is added with the objective of enhancing service, it is imperative that management objectively assess whether the achieved goal was commensurate with the investment in human capital.

JOB ANALYSIS

Part of the planning process, a **job analysis** is necessary before the recruiting and selection processes can begin. Job analysis is the process of gathering, analyzing, and synthesizing information about jobs. As depicted in Figure 8.1, an accurate job analysis is critical for all human-resource management decision processes including selection, orientation, training, appraising performance, compensation, job design, and maintaining legal compliance.

A job analysis consists primarily of a job description and a job specification. The steps involved in a job analysis include collecting information using the appropriate techniques and summarizing the information, including the formulation of a job description and a job specification. The job-analysis process eventually leads to the development of a job design.

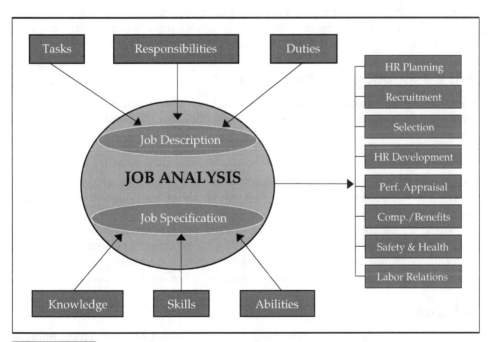

FIGURE 8.1 The Role of Job Analysis in Human-Resource Management.

Data Collection

After establishing the appropriate sources for data on such factors as current employees, related operations, or businesses with a structure similar to that of the particular on-site operation (e.g., that of a chain restaurant), management must decide on the method for collecting the information. This is usually accomplished through observations, interviews, questionnaires, or a combination of these approaches.

Observations are often made using a **critical-incident technique** in which supervisors and employees generate behavioral incidents to measure job performance. Typically, this involves identifying basic dimensions or simple aspects of performance relative to the job. Examples include placing orders, receiving food deliveries, and so on. Next, a list of critical incidents of behavior is generated that pertains to the dimensions. This serves as the baseline indicating what is highly desirable, acceptable, or unacceptable for the particular job dimension. Then the critical incidents are reviewed to ensure that the observed incidents are commonly viewed by different employees. Finally, the critical incidents are assigned effectiveness values that equate to the particular dimension.

Data collected during interviews are typically generated from structured interview questions. Examples include "Describe your typical day" and "What kinds of formal and informal training have you received relative to your position?" Such questions are geared toward collecting information about mental processes, work output, and relationships with others, among other things, that might impact the job.

Questionnaires provide another way of collecting information as part of the job-analysis process. A typical job-analysis questionnaire is similar to an interview in that it collects information pertaining to certain job dimensions such as work context, decision-making processes, and reasoning processes, as well as physical activities. Such surveys are highly structured and, as with any survey, deserve considerable testing for validity and reliability before being used.

Job Description and Specification

A **job description** is a thorough, written summary of a job. It includes the position's title, a summary description, and a list of essential tasks, duties, and responsibilities. This might require differentiation between task-oriented and people-oriented aspects of the job.

In contrast, a **job specification** is a written summary of the knowledge, skills, and other characteristics necessary for effective job performance. The job specification should also include specifics on the minimum qualifications, such as a college degree or culinary degree. In some cases, key elements of the job specification are embedded in the job description.

Without an accurate job description that expresses the frequency and importance of tasks and responsibilities, on-site foodservice managers have no basis for assessing performance. Similarly, without thorough job specifications, managers find it difficult to hire appropriately. This makes the job analysis all the more important.

Job Design

The final reason for performing an in-depth job analysis is that it leads to a thoughtful **job design.** Job design is the process of organizing work into the tasks required to perform a specific job. Four basic techniques comprise job design, all of which flow from the job analysis. The first is **job simplification,** in which a job is broken down into the smallest components (e.g., time and motion analysis). Another approach is **job rotation,** in which a foodservice employee rotates among different jobs that are similar in scope and responsibility. This technique is a precursor to more thorough cross-training, which is discussed in detail in Chapter 13. The next technique is **job enlargement,** the process of broadening existing jobs by adding other related tasks. This is often referred to as "horizontal job expansion." The last technique is **job enrichment;** this entails vertical job expansion, whereby loosely related responsibilities are added to an existing job.

Many on-site operators do not recognize the value of job design, arguing that working in foodservice requires everyone to wear a variety of hats during the day-to-day routine. More savvy managers who use job analysis and job design to develop **workflows,** however, understand that these human-resource processes lead directly to optimized operations. Workflows, in particular, which define how work is organized to meet production and service needs, are extremely useful in training and assessment. Because they are so much more detailed than job descriptions, workflows are also useful in helping employees understand the importance of their jobs within the framework of the entire organization.

RECRUITMENT

Following human-resource planning and job analysis, **recruitment** is vital to ensuring that the demand for quality employees is met. Recruitment is the process of attracting job candidates who have the abilities and attitudes needed for a specific position. The elements of effective recruiting include understanding internal and external factors, utilizing appropriate methods, providing realistic job previews, and evaluating the recruiting methods used by the management team.

Internal and External Factors

Internal factors begin with the demand for employees. As discussed earlier, demand should be linked to the overall strategy of the operation. From a recruitment perspective, demand should be calculated for various time horizons, including immediate, short-term, and long-term needs.

The next internal factor involves organizational policies and practices. For example, does the host organization require all positions to be posted for a set amount of time before the positions can be filled? What about internal hires? Many companies maintain a policy that preference should be given to internal promotions. Another practice pertains to hiring relatives. Some organizations endorse hiring relatives but specify that spouses should not work in the same area or department (a policy for which the company must have a good rationale).

Finally, factors involving organizational culture and image are vital. Take Disney, for example. The organization is very clear about grooming standards. Some foodservice operations have a culture that requires everyone to adopt a service-minded approach to the business. Failure to give this factor its due during the recruitment process will lead to unnecessary turnover and dissatisfaction.

External factors are generally more apparent. Labor-market conditions, for example, have a substantial effect on recruiting (and are directly linked to the issue of supply discussed earlier). Union restrictions are another external factor with great impact. Finally, any recruiting effort must involve a thorough understanding of governmental and regulatory restrictions and applicable laws, including such issues as employee eligibility.

Recruiting Methods

Traditional recruiting methods were once very effective. It was common for an on-site operator to place an advertisement in the local paper and, as a result, have a number of individuals respond. Similar approaches, such as ads in trade and professional publications, were equally fruitful. Finally, asking employees for referrals was a common and useful practice. As the efficacy of these methods declined, however, operators began using other methods common in other sectors such as employment agencies, search firms, open houses, job fairs, summer internship programs, and participation in database-centered endeavors such as those used by many universities.

While some of these methods are still useful in certain instances, they typically produce very small pools of applicants. As a result, various nontraditional methods have emerged. The most common of these is point-of-sale messages. The problem today is that customers are so familiar with such messages in quick-service restaurants that they no longer stand out as unique communication vehicles.

Other nontraditional methods that apply in certain instances are talent scout cards and partnerships. Talent scout cards are useful in recruiting employees from other operations such as restaurants and hotels. When the primary criterion is finding employees with a service mentality, this approach is particularly useful because the recruiter has already witnessed these characteristics during the service exchange with the potential employee. Similarly, partnerships with other companies and schools can be very useful. For example, many colleges and universities—as well as a growing number of high schools—offer hospitality-management programs. Partnering with such institutions can produce a rich pool of potential employees.

Job Previews

The following parable provides a good introduction to **job previews.**

A HUMAN RESOURCES PARABLE
The Job Preview

Once upon a time, a highly successful human-resources executive who had worked for years in the foodservice industry was tragically hit by a bus and died. "Welcome to heaven," said St. Peter. "Given your background, we felt it was appropriate for you to choose where you would prefer to spend eternity. So, we're going to let you have a day in hell and a day in heaven; then you may choose."

And with that, St. Peter put the executive in an elevator and it went down, down, down to hell.

The doors opened, and she found herself stepping onto the putting green of a beautiful golf course. In the distance was a country club, and standing in front of her were all her friends, including many fellow executives whom she had worked with during her career. Everyone was dressed in evening gowns and cheering for her. They ran up and kissed her on both cheeks, and they talked about old times. They played an excellent round of golf and at night went to the country club, where she enjoyed an excellent steak and lobster dinner. She met the Devil, who seemed like a nice guy—nothing like she expected. She had a great time telling jokes and dancing. She was having such a good time that before she knew it, it was time to leave. Everybody gave her a hug and waved goodbye as she got on the elevator.

The elevator went up, delivering her back to the pearly gates. Again, St. Peter welcomed her. "Now it's time to spend a day in heaven," he said. With that, she spent the next 24 hours lounging around on clouds, singing, and playing the harp. She had a great time, and before she knew it, her 24 hours were over.

"So, you've spent a day in hell and you've spent a day in heaven. Now you must choose your eternity," he said.

The woman paused for a second and then replied, "Well, I never thought I'd say this. I mean, heaven has been really great and all, but I think I had a better time in hell."

Reluctantly, St. Peter escorted her to the elevator and again she went down, down, down, back to hell. When the doors of the elevator opened, she found herself standing in a desolate wasteland covered in garbage and filth. Her friends were dressed in rags and were picking up the garbage and putting it in sacks.

The Devil came up to her and put his arm around her. "I don't understand," stammered the woman. "Yesterday I was here, and there was a golf course and a country club, and we ate lobster and we danced and had a great time. Now there is only desolation and despair, and all my friends look miserable."

The Devil looked at her and smiled. "Yesterday we were recruiting you; today you're staff."

The lesson here should be obvious. Despite the common knowledge of long hours and less than heavenly challenges, some foodservice managers sugarcoat things during the recruitment process. Such inflated job previews achieve the objective of maximizing the number of people who accept job offers. The experience once on the job, however, creates dissatisfaction and frustration for employees. Thus, realistic job previews, which set accurate job expectations, are preferable. These previews may result in fewer candidates, but the retention of these individuals will be far greater.

Evaluation

Just as managers evaluate sales, they must also evaluate recruitment efforts to ensure that the investment of time and money is producing worthwhile results. A number of ratios are useful tools in making such evaluations. For example, **selection ratios** (the percentage of individuals hired from the overall selection pool) and **yield ratios** (the percentage of individuals hired from specific recruitment sources) are good objective metrics. Measures pertaining to the quality of applicants, although sometimes difficult to determine objectively, offer another good approach.

Using financial data, calculation of the cost per applicant hired provides a dollar amount that often surprises managers. The inputs should include the proportional salaries of everyone involved in the recruiting process, marketing materials, cost of advertisements, and travel expenses. A related measure is elapsed time between the identification of the demand and the eventual recruitment. In many circumstances, this can be lengthy, resulting in headaches for everyone.

The final means of evaluating the recruiting process is adherence to equal employment opportunity (EEO) and affirmative action goals. Since EEO laws pertain to everyone, adherence is a legal issue as well as an ethical one. Failure to adhere to such laws can result in hefty costs. Affirmative action goals, set by the individual organization, also are often the result of companywide initiatives. Measures determining adherence to these goals provide useful data and may equip managers with the tools they need to minimize unnecessary legal liability.

SELECTION

Selection—the process of choosing individuals who have relevant qualifications—is a large, complex activity, one to which entire textbooks are devoted. This is not surprising, since the selection process ultimately determines the overall quality of an organization's human resources. This is particularly important in an industry that is service and people centered.

The selection process also entails **placement,** or matching the right person with the right job. Estimates suggest that ineffective hiring can cost employers three to five times an employee's annual salary. Yet mistakes—from the smallest, seemingly innocuous to the largest, some with huge legal ramifications—are commonplace.

That being said, one aspect of the selection process has particular significance to on-site managers—the interview. Throughout most of recent history, hospitality managers have relied predominantly on seat-of-the-pants approaches to interviews and the resulting selection process. Perhaps this is because of the business's service nature that rests largely on interpersonal dealings, resulting in managers who need to trust their gut instincts when dealing with unique situations. Or maybe managers assume that since they can manage a diverse workforce well, they can also predict who will perform well without the aid of pertinent, objective information. Solutions to some of these problems entail using effective interview techniques and maintaining an awareness of common mistakes associated with the interview process that lead to bad decisions (for example, see Figure 8.2).

Interviewing Basics

It is easy to assume that simply sitting and talking with a job applicant qualifies as an interview. Granted, such a discussion can yield information about someone, but it in no way resembles the makings of a good interview. Interviewing requires planning, effective questioning, and active listening.

Interview questions and formats vary with the type and nature of the on-site operation. In any case, however, the questions should be linked directly to the job analysis. There is no point in asking questions that are beyond the scope of the position the operator is looking to fill unless there is a distinct reason for collecting such information.

There are several types of interviews, each of which applies to foodservice. The first is the **structured interview,** which uses the same set of standardized questions

FIGURE 8.2 Interviewing Issues. © Tribune Media Services, Inc. All Rights Reserved. Reprinted with permission.

for every applicant. Such questions help the interviewer get an idea of the caliber of the applicant. The next type is the **situational interview,** which entails using structured questions to determine how applicants might handle specific job situations. In some cases, interviewers code responses to such questions as "good," "average," or "poor," thereby allowing for quantitative comparisons.

Interviews that integrate **behavioral questions** ask applicants to give specific examples of how they have performed in the face of a problem in the past. An example is: "Tell me about your most frustrating encounter with a customer. How did you handle the situation?"

Panel interviews involve several interviewers meeting with the applicant at the same time. This allows all interviewers to hear the same response; comparisons and potential biases of individual interviewers are thereby easily identified. But this approach sometimes causes discomfort on the part of the applicant, since the perceived pressure is greater. The approach is often useful, however, for jobs that entail front-of-the-house functions.

The best interview integrates a combination of these approaches tailored to the position that needs filling. Some situations will also dictate what type of interview is reasonable. The key is to conduct the interview according to a plan, not as dictated by lack of preparation.

Common Interview Mistakes

The most common—and potentially most costly—mistake is failing to develop and document objective job criteria, which makes it impossible to discover during the interview whether the candidate is qualified for the position. This is particularly important in light of EEO laws. Related mistakes include using unstructured interviews and tailoring interviews to the candidate as opposed to the position. In other words, in interviewing, a manager must use the same methodology for all applicants for a specific position.

Other related mistakes can result in bias. Inexperienced interviewers tend to select applicants whom they perceive are similar to them or share their interests. This is referred to as **similarity error.** A similar problem is the **halo effect,** which commonly occurs when an interviewer allows a prominent characteristic such as a degree from a specific university to disproportionately but positively influence his or her decision. The opposite is known as the **devil's horns,** in which an interviewer places undue negative weight on a single attribute or accomplishment.

Contrast error results when the interviewer compares only those persons in the current interview pool. Another mistake is when an interviewer fails to remember different applicants and instead permits the first or last applicant to influence the decision unfairly, known respectively as **primacy** and **recency bias.** Finally, the interviewer has a responsibility, both to the applicant and to the organization, to listen actively to every applicant. While this is sometimes extremely difficult, it is the cornerstone of equity in the selection process.

TRAINING AND DEVELOPMENT

While finding quality employees is pivotal to an on-site foodservice operation's survival, training and development arguably play a greater role in achieving long-term financial success. **Training** is the process by which employees acquire capabilities that contribute to the overall success of the on-site operation. **Development** pertains to efforts intended to enlarge employees' skillsets to better handle current and future challenges associated with on-site foodservice.

Training is usually targeted at the current job and is expected to produce immediate improvement. Training also is typically focused on individuals with the explicit objective of remedying a skill deficit. Development has a more long-term focus; it prepares individuals and groups for demands they will likely encounter in the future. Not surprisingly, both of these processes entail very specific key steps.

Training

The critical success factors in any training program include managerial involvement, alignment and integration of the operation's systems, consideration of unambiguous job characteristics, and sensitivity to and awareness of social dynamics. On a more micro level, components of effective training programs include a needs analysis, clear objectives, thoughtful program implementation, and objective evaluation.

A **needs analysis** should be performed along three dimensions. At the organizational level, management determines *where* training is needed. Relevant information may include staffing patterns, demand forecasts, expected changes in equipment or technology, and so on. *What* job-related tasks, duties, or responsibilities require training must also be decided. This may include a determination of what tasks involve customer interaction or whether a position is more technically oriented. Finally, *who* needs the training must be determined. For example, is training needed for current or new employees? Important concerns include education and previous training, work experience, current work-performance indicators, and career interests.

Training objectives of the program or initiative must be clearly articulated. It is insufficient simply to state that trainees will understand the basics of dining-room operations. A better, clearer objective would read: "Trainees will be able to identify the three primary ways to exceed guest expectations."

Program implementation usually necessitates one of five basic delivery methods. The first is the common lecture, which is efficient for disseminating basic information to large groups but ignores individual differences among trainees and makes learning job-related skills difficult. Second, discussion and demonstration are good for technical, hands-on tasks and provide an opportunity for trainees to discuss what they are learning. This approach is effective, however, only when used with small groups and requires a trainer who is highly proficient in performing the specified tasks.

Third, integrating a combination of audio and visual materials is useful when there is a need to illustrate task sequences and is good for training many employees at different sites; it also is excellent for tasks that are difficult or costly to demonstrate. On the other hand, using audiovisual materials can make it difficult to teach some tasks, duties, and responsibilities and may be very costly to produce. A fourth, much cheaper method is case studies, which are excellent for developing cognitive skills. The challenge is to prevent trainees from assuming that there is one correct answer (thereby undermining the intent of the training method).

Fifth and final, role-plays, games, and simulations provide excellent opportunities to practice and demonstrate knowledge and skills; this approach also allows for immediate feedback. Such methods, however, lack the incentive of true risk and may be perceived as more enjoyable than instructional. They also require a high level of proficiency on the part of the trainer. Finally, role-plays, games, and simulations can generate unhealthy competition among employees.

The best approach typically involves four steps and may entail any combination of the aforementioned delivery methods. The steps are (1) preparation, or telling; (2) demonstration, such as showing how something is done; (3) practice on the part of trainees; and (4) follow-up on the lesson, usually including a review.

In order to ensure that the training is worthwhile and beneficial to both the employees and the organization, **objective evaluation** using detailed criteria is important. Such criteria fall into four categories: reactions, learning and knowledge, behavior and performance, and results.

Immediate reactions fall into two categories, affective and utility. Affective reactions include those that relate an emotional response to the experience, such as "I am pleased that I attended this training program." Utility reactions draw a link between the experience and the work environment, such as "This training will help me perform better." Reactions are a useful source of feedback for trainers. It is important that while training should generally be enjoyable, utility is the most important benefit.

The second means of evaluating the training program is testing trainees' resulting knowledge and learning. For example, asking employees to reiterate definitions or desired outcomes can serve as a good measure. If a trainee is unable to provide sufficient answers, this is a likely indication that the objectives of the training program have not been met.

Tests of behavior and performance provide another dimension of objective evaluation. For example, to what extent does the newly trained worker perform differently? Management can make such determinations more easily for certain positions (say, for a cashier who just underwent training on the point-of-sale system) than others. Nonetheless, this is often an invaluable criterion of training effectiveness.

The final objective-evaluation criterion is a comparison of pre- and posttraining differences in customer-service ratings, formal and informal grievances, turnover, absenteeism, and so on. These statistics, when aligned with the training initiatives, serve as concrete measures of whether the training program is delivering value to the organization. Training that fails to produce positive results in this area represents poorly allocated resources.

Development

Employee development, like training programs, must begin with an analysis of (1) the organization and (2) the individual. Failure to focus on both is a recipe for the long-term failure of any development program. Furthermore, it is important to note that development is not reserved for management. Line employees can benefit greatly from development and can bring considerable value to the organization when such development is appropriately delivered.

The first step, then, in launching an employee-development initiative is to identify the present and future needs of the company. This is often tied to the strategic plan and requires considerable interplay between operations managers and human-resource managers. As discussed earlier, this also entails the determination of demand as applied to human capital.

With the organization's needs clearly articulated, the two most important aspects of employee development are **development-needs assessment** and **succession planning.** Both the individual and management should perform the development-needs assessment. The goal of the assessment process is to identify strengths and areas of opportunity for future development for each employee. Methods vary by organization, but the most popular are assessment centers, psychological and performance tests, and performance appraisals.

Succession planning is similar to playing checkers. For example, it involves planning where different employees might fit in the organization in the future, much as a checkers player tries to anticipate the next two or three moves. Succession planning integrates the findings of the development-needs assessment and integrates expectations of how employees will progress given adequate development opportunities. In many instances, this translates into identifying line employees who have the potential for management opportunities. Succession planning can be tricky, particularly in larger on-site operations or in multiunit settings. Nonetheless, thoughtful succession

BEST PRACTICES IN ACTION

TRAINING AND DEVELOPMENT AT UCLA MEDICAL CENTER (LOS ANGELES, CALIFORNIA)

UCLA Medical Center has been known for its pioneering technological contributions since it opened its doors in 1955. With over 600 beds, over 1,000 physicians, and 3,500 nurses, therapists, technologists, and support personnel, the healthcare leader provides services to some 300,000 people from different parts of the world. The Level 1 trauma center, the latest diagnostic technology, and the eye research center are just some of the features that stand out.

Until recently, however, foodservice was dwarfed in terms of efficiency, level of service, and fiscal prudence compared with the other departments within the Medical Center. The problems ranged from a demoralized staff to alleged criminal activity associated with disappearing food products, with annual losses exceeding $2.5 million. Patient satisfaction had bottomed out, to stay nothing of the reputation of the food served in the cafeteria. Simply stated, the supervisory staff was untrained to meet the challenges of running such a large operation; the line staff members were even more devoid of the training and development they needed.

Today, things have changed. According to the new management regime, what has transpired is a "turnaround through training." The driving theme behind the comprehensive training initiative is a holistic view of the operation that includes—and emphasizes the importance of—every employee. For supervisory staff members, this has translated into weekly training sessions spanning 18 months. In essence, the program takes a "train-the-trainer" approach, in which supervisors are equipped with both the skills they need and the prowess for disseminating information, tools, and approaches to their respective employees.

Who attends these training sessions? The 17 supervisors within the department, as well as those who have been identified as possessing supervisory potential (as part of a formal succession-planning system). Thus, the program is designed to foster learning and to underscore the importance of development.

Perhaps the most impressive aspect of the training is that it is not specifically skill based. Instead, it focuses on behavioral and attitudinal factors that have greater overall relevance to all the functions of the department. In addition, the training embraces the diversity of the staff (whose members represent some 30 countries) and customers, with the desired outcome being optimization.

While the approach—and considerable resources dedicated to training—is inspiring, the results are even more noteworthy. Following the bulk of the initial training program (the training is now an ongoing process), sales in the cafeteria increased by over $100,000 per month and annual costs were reduced by approximately $11 million. Catering sales went from less than $80,000 per year to $1.2 million.

But what about the level of service? The accomplishments speak for themselves. For example, a couple of years ago, the department was awarded the prestigious Silver Plate award in recognition of the service as well as the quality of the food offered by the Medical Center to its many constituents.

Today, the foodservice department is one of the jewels in the UCLA crown and is no longer an area that the organization needs to restrict to the basement.

planning provides opportunities for employees to grow while remaining in the organization, one of the keys of retention management.

The importance of training and development cannot be overstated. One management team that has realized this and has acted on it, with impressive results, is featured as a best practice.

RETENTION MANAGEMENT

As every foodservice manager knows, the biggest challenge following the selection of good employees is retaining them. Underscoring the challenge is the reality that employee turnover and foodservice seem bound together. Some say that this is inevitable and argue that managers must learn to deal with it as with any other operational realities, such as ever-rising prices. The best managers, however, know that turnover is a symptom of deeper organizational troubles.

The issue that makes turnover so horrific and justifies strategies geared to enhancing retention is the cost associated with losing employees. Some research suggests that the cost of turnover averages $5,000 per hourly employee, but other findings more than double that number. These astronomical costs do not seem so inconceivable when one considers what turnover entails. For example, there are costs associated with the actual separation, such as severance pay and paperwork processing, as well as the lost productivity of employees who are on their way out the door. The costs of recruiting and selecting are enormous, including advertising costs, management time spent on interviews, and the time and expense of background and reference checks. Then there is the cost of training once the new employee is hired. Finally, there is the sizable cost associated with loss of productivity stemming from the unplanned vacancy.[2]

In order to combat turnover and the issues that lie beneath it, managers must first know how to determine the rate of turnover. With this information in hand, they can then consider some of the key methods of managing retention.

Determining Turnover Rates

Unfortunately, managers do not always calculate turnover the same way, which makes comparisons somewhat difficult. Even worse, some managers do not calculate turnover consistently, which makes any determinations following the calculation suspect.

There are three general ways to calculate turnover. The first, advocated by the United States Department of Labor, is as follows:

$$\frac{\text{Number of employee separations during the period}}{\text{Total number of employees at the midpoint of the period}} \times 100 = \text{Turnover rate}$$

This method is inadequate because it assumes that employee separations will occur at equal intervals throughout the period. For example, assume that an on-site operation has 28 employees at the beginning of the month (the period for which the turnover statistic is needed). If five employees depart before the middle of the month, the turnover rate, as calculated with this formula, is just under 22 percent. If only one employee leaves at the beginning of the month and four leave near the end, however, the rate as calculated is less than 19 percent.

The second method allows for the often subjective determination between the "desired turnover" of undesirable employees and the "undesired turnover" of desirable employees. This approach then treats the resulting statistic as the "undesired turnover statistic." The calculation is performed as follows:

$$\frac{\text{Number of separations} - \text{desired separations}}{\text{Average number of employees during the period}} \times 100 = \text{Undesired turnover rate}$$

This calculation appropriately takes the average number of employees during the period (as shown in the denominator). In essence, however, this method is useful only to justify some elements of the turnover problem. The truth is, turnover is a bad thing. Yes, poor hiring decisions are made, which then result in turnover. Nonetheless, this turnover reflects problems in the human-resource management process and cannot be glossed over.

The final method is the easiest and offers the most utility. The calculation is:

$$\frac{\text{Number of employee separations during the period}}{\text{Average number of employees during the period}} \times 100 = \text{Turnover rate}$$

This statistic is not subject to interpretation and also allows for intra- and interunit analysis (in the case of multiunit on-site operations).

Retention-Management Methods

Retention-management methods are reasonably straightforward, albeit complex to implement effectively. In essence, they apply to the major topics already discussed in this chapter: recruiting, selection, training, and development. Poor recruiting leads directly to poor selection. A simple example is a case where recruiters fail to tap into the appropriate labor pools. Selection, as discussed earlier, is also a difficult process even when recruiting produces a good group of potential employees. The associated challenges are underscored by the many considerations necessary during just one part of the process discussed earlier: the interview.

Next, inadequate training equates to inadequate performance on the job. Inadequate performance, then, leads directly to dissatisfaction on the part of the employee and the manager—and, more than likely, the customer. Training issues more than any others impact both the internal and external constituencies, even when the best employees are placed in the right positions.

Finally, development is an area that is vital to long-term retention. Employees must perceive that there are opportunities for them to grow with the organization and that the organization is willing to dedicate resources to their development. Development also speaks directly to maximizing productivity, the central topic of the following chapter.

CHAPTER SUMMARY

Planning is the first human-resource management process and leads to ensuring that an on-site operation is staffed appropriately. It involves environmental scanning, assessing demand and supply, designing and implementing associated programs, and evaluating the outcomes of these programs.

The next process is the completion of a job analysis, by position, for the operation. An accurate job analysis is critical for all human-resource management issues including selection, orientation, training, appraising performance, compensation, and legal compliance. This entails collecting data, developing job descriptions and specifications, and, ultimately, developing appropriate job designs.

Recruitment is critical to ensuring that the demand for quality employees is met. Effective recruitment involves evaluating internal and external factors, employing appropriate recruiting methods, integrating accurate job previews, and evaluating the overall process. Selection is the flip side of the recruiting coin and is arguably a more complex process, which includes accurate placement of recruits. Despite its complexity, the biggest selection-related hurdle most commonly encountered by on-site foodservice managers involves the interview process. Several interview formats can be effective, but each requires appropriate planning. The key is to avoid common yet costly mistakes that invariably lead to employee dissatisfaction, management frustration, and turnover.

Training and development should be front and center in all foodservice managers' minds. The training process requires a needs analysis, the development of training objectives, thoughtful program implementation, and objective evaluation. Development requires the identification of the organization's needs, followed by a development-needs assessment and succession plan specific to the operation's employees.

Retention management is, ultimately, the basis for all human-resource-related activities. In order to be effective, retention management first requires accurate calculation of turnover. From there, appropriate retention-management methods can be employed, each of which entails various aspects of the aforementioned key topics: recruiting, selection, training, and development.

KEY TERMS

planning process	workflows	devil's horns
environmental scanning	recruitment	contrast error
expert-estimate technique	job previews	primacy bias
trend-projection technique	selection ratios	recency bias
job analysis	yield ratios	training
critical-incident technique	selection	development
job description	placement	needs analysis
job specification	structured interview	training objectives
job design	situational interview	program implementation
job simplification	behavioral questions	objective evaluation
job rotation	panel interviews	development-needs
job enlargement	similarity error	assessment
job enrichment	halo effect	succession planning

REVIEW AND DISCUSSION QUESTIONS

1. What does environmental scanning entail? What are some applications where it might be useful?
2. Suppose that you had to forecast the demand for an on-site operation. What technique would you use, and what would the process entail?
3. Name three ways in which job descriptions can be useful for an operator.
4. How can workflows help a manager? Are they worth the time and trouble?

5. Realistic job previews may discourage very qualified candidates from accepting some foodservice jobs. Is this a bad thing?

6. In what ways might behavioral questions allow an interviewer to gain information that might otherwise be missed during an interview?

7. A job applicant graduated from the same college as you (only much more recently). Given the discussion, how might this affect your decision?

8. Should development be targeted more at managers or at line employees? Explain your answer.

9. An on-site operation boasts that line-employee turnover was only 50 percent for the past year. Yet, in looking at the numbers, you calculate turnover to be much closer to 100 percent. How is this possible?

END NOTES

[1] Of course, other methods exist for human-resource-related forecasting. A good source for these is Rothwell, W. J., & Kazanas, H. C. (1988). *Strategic human resources planning and management.* Englewood Cliffs, NJ: Prentice Hall.

[2] There are several good articles that talk about cost in more detail. Two of these are Hinkin, T. R., & Tracey, J. B. (2000). The cost of turnover: Putting a price on the learning curve. *Cornell Hotel and Restaurant Administration Quarterly, 41*(3), 14–21; and Woods, R. H., & Macaulay, J. F. (1991). R$_x$ for turnover: Retention programs that work. *Cornell Hotel and Restaurant Administration Quarterly, 30*(1), 79–90.

CHAPTER 9

PRODUCTIVITY

This chapter focuses on productivity, a topic often discussed but not so often clearly defined or quantified in foodservice-management circles. To this end, the chapter begins with a definition of productivity as it applies to on-site foodservice along with the core issues surrounding labor productivity. The discussion then turns to methods of measurement. As discussed in previous chapters, any management tool's value is nested in the ability to measure its effectiveness.

With a clear definition and a segment-specific measure in hand, the chapter then moves on to issues of enhancement. Enhancing productivity is critical to any foodservice operation, but it is particularly germane to on-site foodservice given the emphasis on doing more with less. The best practice in this chapter offers a rich example of a program aimed at just that.[1]

DEFINITIONS AND CORE ISSUES

In its most general application, **productivity** is a performance measure and can be defined as *the effective use of resources to achieve operational goals.* On the basis of this definition, the basic industrial model defines productivity as output divided by input. Extending this notion, productivity can be represented by the following equation:

$$\frac{\text{Goods} + \text{Services}}{\text{Labor} + \text{Materials} + \text{Energy} + \text{Capital}} = \text{Productivity}$$

Productivity can also be expressed as either a **partial-factor statistic** or a **multiple partial-factor statistic** by selecting at least one of the listed variables from both the numerator and the denominator. Partial-factor productivity statistics, however, may not be good indicators of overall performance since they serve only as measures of isolated aspects of an operation.[2]

The problem arises when managers analyze these partial-factor productivity measures as indicators of overall operational performance without considering the effects of related variables. Moreover, partial-factor productivity is often used as a surrogate for profitability, since it follows, ostensibly, that optimal utilization of labor, materials, energy, or capital would equate to increased profits. Effective treatment of any one of these, however, does not ensure overall performance. For example, if we define productivity as the ratio of the number of guests served to the number of food servers, and then apply this measure to an upscale restaurant, a targeted effort to maximize productivity would likely have an adverse effect on long-term profits.

Similarly, if a kitchen manager replaces labor-intensive food items with prepared products in order to decrease labor costs, the findings may indicate greater productivity in terms of total labor-dollar expenditures. However, the cost of the prepared items may exceed the reduced cost of labor. As a result, labor productivity is improved but the financial viability of the operation becomes questionable. In this case, if productivity had been measured by dividing sales by the combined cost of labor and food, the wisdom of the operational change, at least in short-term financial terms, would have been accurately deduced.

MEASUREMENT AND ANALYSIS

How do we measure productivity? Some partial-factor productivity statistics can be very meaningful and can be useful as indicators of operational performance, most commonly in labor management. Poorly formulated or misunderstood measures can be damaging, however, and, if relied on as the primary indicator of performance, disastrous. The key is to select outputs and inputs that indicate which operational areas are most in need of scrutiny. When total operationwide productivity is the target of analysis, aggregate measures that integrate the majority of operational inputs are appropriate.

Differing Measures

The first measure of productivity was applied in B&I, and was adopted from the quick-service-restaurant (QSR) segment. As such, the focus was on sales and labor, with the measure being **sales per labor hour,** since food cost varied little owing to the segment's heavy emphasis on product standardization. This partial-factor statistic was very useful in fast-food operations, particularly because labor outpaced food cost as the largest expense item.

As originally applied in B&I, sales were treated as total sales net of sales tax. Labor hours were calculated using productive hours only, which include regular time, overtime, on-call hours (if worked), and hours spent in training and orientation. Salaried positions were accounted for on the basis of eight hours per day per person (even if a manager worked substantially longer shifts).

Three operational issues that separate B&I operations from their quick-service brethren make this measure less than ideal. First, some B&I operations are subsidized. Thus, sales of one operation cannot be compared with those of another that has a different subsidy or no subsidy built into the price structure. Second, and less apparent, is the existence of satellite outlets. Satellites, which range from small coffee kiosks to fully functional eateries, are generally considered financially viable because they maximize distribution and food-production economies. They can, however, wreak havoc on sales-per-labor-hour statistics.

For example, suppose that two B&I operations have average sales of $2,700 during a typical weekday morning, with hours of operation beginning at 6:00 A.M. and ending at 10:00 A.M. Unit 1 is a single eatery with a full staff totaling 82 labor hours and typically has a 31 percent food cost for the morning daypart. Unit 2 consists of a much smaller eatery and a satellite kiosk. The labor for the eatery is only 60 hours; because the kiosk features only coffee and individually wrapped baked goods, it requires only 4 hours of labor. The total labor for Unit 2, then, is 64 hours. The food cost is much higher, however (approximately 44 percent), because of the emphasis on prepackaged goods. On a sales-per-labor-hour basis, Unit 2 is a superstar, with a sales-per-hour productivity statistic of $42.2. Conversely, Unit 1 looks like a dog, with a statistic of $32.9. Obviously, Unit 2 is more "productive," but is it more efficient in terms of profitability?

The third challenge in using this fast-food-borne partial-factor productivity statistic relates to catering activities. Some on-site operators include catering sales in the sales-per-labor-hour productivity measure. Others keep catering sales separate but cannot always separate the labor used for catering activities. For example, when the entrée is prepared en masse both for a catering function and for sales in the main cafeteria, how much of the cook's preparation time should be attributed to the catering activity? Similarly, how should managerial hours be distributed between catering and routine operations when the same manager oversees both?

Catering activities also typically require fewer overall labor hours than do other operations of the on-site entity. As a result, productivity measured by sales per labor hour will be artificially inflated by heavy catering activities. (Again, this is detrimental only if catering as a factor of the productivity indicator is handled differently across units.) Furthermore, if catering activities are not priced properly, increased sales from catering may increase the sales-per-labor-hour statistic but will not enhance the financial performance of the unit.

In the corrections, education, and healthcare on-site foodservice segments, **meals per labor hour** is a common productivity measure. For corrections, this is a reasonable statistic, at least to determine labor productivity, since the number of meals served is a direct function of the number of inmates. Moreover, the number of different types of meals is limited, although this situation is changing today as greater emphasis is placed on meeting prisoners' varying dietary restrictions and preferences.

In the case of schools, however, which are largely government subsidized, this measure is fraught with problems. First, a school meal is typically considered the equivalent of an entrée, two side dishes, and a beverage; à la carte items are not often taken into consideration (although they arguably constitute a small percentage of total sales). Second, subjective meal equivalents are used in tallying sales from business outside the central school. Since operators are increasingly seeking add-on business—from child daycare centers, adult daycare centers, preschools, and private schools, where meals are sold at a reduced cost to reflect the efficiencies attained in volume food production—meals per labor hour can be greatly, albeit artificially, inflated.

Problems with this partial-factor measure in colleges and universities are similar. Subsidies vary widely by institution. In addition, meal-plan configurations are so diverse that any two schools are rarely alike. This is problematic since each meal plan specifies what constitutes a "meal"; some colleges base this definition on a specific dollar amount, while others equate it to preassigned menu combinations.

In healthcare, counting meals—the essential component in calculating meals per labor hour—is surprisingly complex, even more so than in education. The problems with catering are the same as those in B&I. The issue of quantifying patient meals, however, dwarfs such problems. As noted in Chapter 5, rarely do patients eat three meals a day. This is unfortunate, since it would otherwise be easy to calculate meals—simply take the number of patients and multiply by 3. In addition, some patients receive tube feedings, which require varying amounts of time to prepare and deliver (tube feedings can also be very expensive in terms of product cost). Should a tube feeding count as a meal?

Other problems associated with accounting for patient meals include those created by the diverse patient mix. Pediatric patients, for example, typically eat smaller meals than their adult counterparts. At the other end of the continuum are psychiatric patients, some of whom are in recovery from addiction or related illnesses and may require double portions at every meal. Should a tray for pediatrics with a three-ounce hamburger, two ounces of potatoes, and juice be considered equivalent to a tray sent to a psychiatric wing with two servings of lasagna, garlic toast, two salads, two large milks, and two desserts? These same differences are also seen in the consumption habits of outpatients; one might reach satiety with a cup of broth, while another enjoys a multi-item meal.

Three other important components of patient meals also strongly influence the meals-per-labor-hour determination. Floor stocks, such as juices, coffee, and crackers, which are maintained at par levels on various floors, augment a patient's intake of calories. These must be routinely replenished, which equates to labor hours. Dietary supplements, used in a variety of healthcare settings to dramatically boost patients' caloric intake, are difficult to translate into meals yet are a real cost to the department (with some costing only pennies per serving and others costing several dollars). Finally, between-meal snacks, commonly referred to as "nourishments" (or as "10/2/8s" to reflect the times when they are usually served), are a big part of many patients' meal service.

Many healthcare foodservice managers have addressed some of these issues through the use of an **equivalent meal value** (EMV) equal to the cost of the raw food used for a typical midday meal. (Most operators consider a typical meal as an entrée, two side dishes, a beverage, and a dessert.) While the EMV varies according to the user, most agree that it is somewhere around $1.50. This factor can be used for patients, catering, and public eateries.

The obvious problem with using an EMV is that no consideration is given to variations in menu mix, positioning of the institution, or sales strategies. Also, the applicability of such an approach is predicated on accurate food-cost percentages, which, as discussed in Chapter 7, may be inaccurate without sufficient controls in place. Finally, an EMV's appropriateness is particularly questionable when applied to catering. Is a lobster dinner prepared and served to 12 board members the same as 65 boxed lunches served at an orientation meeting? Assuming the raw food cost was equal for the two functions, these two events would equate to the same number of meals when using an EMV. Is that accurate?

A minority of operators have combatted some of the problems associated with using an EMV by replacing it with a **floating equivalent-meal-price factor** (EMP). An EMP is the average price of a complete meal specific to the daypart, catered event, or patient meal. To be effective, the EMP must be calculated repeatedly. The frequency of such calculations is dictated by variations in sales, catered events, and patient eating patterns. Thus, an operator will very likely use several different EMPs during any one day, with each corresponding to the daypart and outlet. Without fairly sophisticated financial tracking systems, such a task is extremely burdensome and may require so many management hours that it in itself saps the operation's productivity.

Before moving on, it is important to note that single-factor productivity statistics can serve a purpose in certain situations, such as **intertemporal analysis** of a single aspect of the on-site operation (which involves comparing the same statistic over time). Such analyses, using meals per labor hour, also offer the advantage that no adjustments are needed for inflation, assuming that the EMV is adjusted accordingly each year. Intertemporal analysis using dollar-based statistics (such as those pertaining to labor costs), however, must be adjusted to reflect constant-dollar values.

A More Robust Approach

Expanding the number of factors in evaluating productivity substantially increases the utility of the measure. Food cost is a critical element here, since it often has a reciprocal effect on labor productivity and always has an effect on the price-value equation. In addition, labor cost (as opposed simply to hours) can be more useful since hours worked do not necessarily correspond proportionately to labor costs.

Perhaps the factor that is most often neglected is capital improvements, which are usually allocated to the foodservice department for the corresponding depreciation period. In a contract environment, these often take the form of **amortized leasehold improvements.** As the term implies, amortized leasehold improvements spread the costs of leasehold improvements made in the foodservice area over a multiyear period, which typically corresponds to the length of the contract. In either case, this is a real expense. Moreover, such improvements should make a favorable impact on sales, or on some facet of productivity, and should be included when measuring it.

Referring back to the example of breakfast sales at the two hypothetical B&I units, we see it was noted that Unit 2 (the one with the kiosk) was more productive in terms of sales per labor hour. It is unknown, however, which unit is truly more productive in terms of efficiency. Using a measure that includes revenue, productive labor cost, cost of goods (which in this example is calculated as food cost only), and apportioned improvements (either capital or leasehold), the answer is clear.

As illustrated in Table 9.1, the expenses for the units are very different and produce distinct productivity statistics. The labor used in Unit 1 is higher both in the number of hours and in the average hourly rate than the labor used in Unit 2. The cost of goods in Unit 1 is, however, dramatically lower than that in Unit 2. Furthermore, it is evident that Unit 2 has a greater proportional improvement cost, which is possibly associated with the purchase price of the kiosk. So, whereas Unit 1 might have been criticized for being less productive according to the single-factor productivity calculation, the addition of other factors in the productivity statistic with revenue in the numerator and the costs in the denominator results in a more accurate picture, one that suggests that Unit 1 is more productive. Granted, this analysis involves only one day's worth of data, but the utility of the measure is apparent.

This example demonstrates that, in terms of the number of partial factors included in the productivity calculation, more is better. Taking it a step further, an optimal

TABLE 9.1

Revenue, Costs, and Corresponding
Productivity Index by Unit

	Unit 1	Unit 2
Revenue	$2,700	$2,700
Productive labor cost	$1,193	$ 832
Food cost	$1,107	$1,458
Proportional improvements	$ 26	$ 54
Productivity index	1.16	1.15

measure of productivity for on-site foodservice is one that is an aggregate, total-factor productivity statistic expressed as

$$\frac{rev_i}{fc_i + lc_i + roe_i + (mp \text{ or } mf)_i + ai_i} = \text{Productivity}$$

where

$$rev_i = \text{Revenue for period } i$$
$$fc_i = \text{Food cost for period } i$$
$$lc_i = \text{Productive labor cost for period } i$$
$$roe_i = \text{Related operating expenses for period } i$$
$$(mp \text{ or } mf)_i = \text{Apportioned minimum profit } or \text{ management fee for period } i$$
$$ai_i = \text{Apportioned investment for period } i$$

In this case, revenue is the sum of sales from all activities, including catering, cafeteria, vending, patient meals, and any other food-related service. Revenue associated with patient meals or other food items is accounted for by meticulous documentation of what leaves the kitchen, including factors discussed earlier such as trays with double portions. Beyond this, only two variables appearing in this equation have not been previously discussed in a specific context. The first is related operating expenses, which include such items as paper goods, cleaning supplies, cooking utensils, uniforms, data and office supplies, merchandising materials, menu or other printing expenses, rent or lease (if applicable), music and entertainment, maintenance and repairs, equipment agreements, licenses, insurance, and depreciation of equipment not included in the apportioned investment amount (a good candidate for depreciation as a related operating expense would be computer hardware)."Minimum profit"refers to a profit margin that must be maintained or a minimum amount that an operation needs to set aside for capital projects. For operations operated by a managed-services company, a"management fee"is the expense paid to the contractor when the contract is structured on a fixed-fee basis.

In practice, this aggregate total-factor statistic is a true indicator of productivity—and performance. A number greater than 1.0 indicates that outputs (sales) exceed inputs (costs); more to the point, it indicates positive performance. While thresholds vary from organization to organization, a reasonably annualized minimum value using this metric is 1.08.

Table 9.2 summarizes a hypothetical statement of operations for the fictional Midtown Medical Center, a 250-bed acute-care facility. Intertemporal analysis shows that the foodservice department has maintained or improved their meals-per-labor-hour statistic during the last few years. Using the fiscal-year data shown, this number

TABLE 9.2

Midtown Medical Center Summary Statement of Operations, Year-End December 31

Revenue		
Patient Meals	$863,596	
Catering Sales	391,500	
Cafeteria Sales	675,328	
Total Revenue		**$1,930,424**
Expenses		
Food Cost		
Patient Meals	$293,600	
Patient-Related Food	34,949	
Catering	145,000	
Cafeteria	243,118	
Total Food Cost	716,667	
Labor		
Salaried	$140,500	
Hourly	666,510	
Payroll Related	235,798	
Total Labor	1,042,808	
Related Operating Expenses	150,636	
Minimum Investment	50,000	
Annualized Leasehold		
Improvement	25,000	
Total Expenses		**1,985,111**
Operating Profit		**($54,687)**
Statistics		
Patient Meals	220,752	
Catering Meals (EMV = $1.46)	99,315	
Cafeteria Meals (EMV = $1.46)	166,519	
Total Meals	486,586	
Productive Labor Hours		
Salaried	8,320	
Hourly	70,512	
Total Hours	78,832	
Meals per Productive		
Labor Hour	**6.17**	
Total-Factor Productivity Index	**0.97**	

is 6.17, which is particularly impressive given the industry average of 3.91 (for similarly sized healthcare providers).

A more accurate productivity analysis using the aggregate productivity statistic suggests that things are not as rosy as one might think. The operation's true productivity, as shown at the bottom of the table, is 0.97, as measured by the total-factor method. This is dramatically less than the target of 1.08. Moreover, the resulting statistic reflects a negative operating profit. Operational analysis would likely uncover one or more issues associated with an unbudgeted increase in labor cost, inappropriate assignment of positions (e.g., having a highly paid, skilled employee performing an entry-level job), unacceptably high food cost, or factors associated with sales, such as price structure or sales mix.

More than any other, this statistic is universally applicable and meaningful as an indicator of performance across units and organizations. It reflects operational efficiencies, such as those gained through maximized employee retention, effective training, accurate market identification, and accurate asset management, any and all of which are achieved through various combinations of economies and operational acumen. It also serves as a flag in the event that a foodservice manager fails to employ judicious business practices.

Some might argue that an aggregate productivity statistic is simply a ratio analysis derived from any income statement whereby sales are divided by expenses and that, as such, a simpler measure of profit (such as a percentage of sales) is more useful. While profit percentages are indeed functional, the utility of the productivity metric is best seen when one is comparing units and discussing the related issues with different constituents. For example, it may be undesirable for an operator to talk about profit with employees, particularly if they do not understand that other expense areas, such as corporate overhead, absorb any operating profit. In a contracted environment, an operator might want to share efficiency statistics with a client but would not be at liberty to share profit information.

Finally, the value of this aggregate productivity statistic is seen in its versatility and applicability in any on-site setting. The relative simplicity of B&I and, to a lesser extent, of the education segment may allow operators to use a multiple-factor statistic that is less encompassing but still valid. In healthcare, however, any measure that does not include all the categories described above simply does not do justice to the complexity of today's healthcare foodservice operation.

ENHANCEMENT

Operating under the maxim that "what gets measured gets done," managers who embrace this somewhat complex approach to quantifying productivity generally achieve better results than those who fail to measure efficiency. In addition, an accurate picture of operational productivity affords managers pathways necessary for improvement. These often take the form of programs and approaches that build on internal-control measures and human-resource initiatives, discussed in previous chapters. Some of these that offer universal appeal and are applicable in most on-site settings include perfect attendance programs, worker safety programs, and recognition programs. Without question, these programs can have rather dramatic effects on efficiency, quality, and other desirable factors such as morale.

Perfect Attendance

Perfect attendance programs strive to reduce unnecessary nonproductive time and, ultimately, to increase overall departmental productivity. Such programs work on the premise that positive behavior should be rewarded and that recognizing behavior such as perfect attendance leads to shared norms that are linked to organizationally desirable outcomes.

In most applications, a perfect attendance program is based on a simple reward-based design with hourly employees in mind. The basic tenet is that any hourly employee who has no unscheduled absences or tardiness during a specified period of time qualifies for an award. The time during which the attendance is measured can be one, three, or six months and can also be measured cumulatively.

In general, on-site operations that have issues with employee attendance use shorter measurement periods for making incremental improvements. Conversely, operations with generally good overall attendance use longer periods (e.g., six months) in order to challenge employees. Perfect attendance programs can be repeated indefinitely, regardless of the measurement period. Furthermore, separate categories can be integrated for employees who demonstrate perfect attendance during consecutive periods. Most operators familiar with such programs suggest that once attendance is under control, a useful approach is to apply the program on a quarterly basis, with special awards for employees with perfect attendance for any two quarters or for one full year.

As for awards, the number and value are generally dictated by the size and budget of the operation. Larger foodservice departments that may have several employees with perfect attendance during a given period may opt to recognize winners with a free pizza lunch along with a prize raffle. For smaller operations, it may be more valuable to present each employee with an individual reward.

BEST PRACTICES IN ACTION

FAIRFAX HOSPITAL'S PERFECT ATTENDANCE PROGRAM (FALLS CHURCH, VIRGINIA)

Fairfax Hospital is a 656-bed regional medical center that serves the metropolitan area of Washington, D.C. Part of the Inova Health System, it is a medical and nursing teaching hospital, and is recognized as a quality healthcare provider. It also shares the problems common to other businesses with an organizational culture that, while intended to focus positively on its people, develops some human-resource-related issues that are problematic. Specifically, the foodservice department once had an issue associated with attendance. It wasn't that the employees weren't good or well trained; it was simply a matter of a culture that accepted "callouts" as a matter of fact. The benefit package for employees allowed for a large number of paid days off and required very little notice if an employee needed a day off. When first instituted, the benefit was intended as a draw for employees. Over time, however, it became one of the facility's most widely abused benefits, particularly in the foodservice area. At one point, in fact, employee schedules were developed with the expectation that there would be several callouts each week.

While challenging in several ways, the problem had become a morale killer. Even for the handful of employees who didn't abuse the policy egregiously, it was difficult to take responsibility for certain tasks when the burden landed entirely in their laps if a fellow employee didn't come to work. It was also tough on the supervisory staff, since they often had to assume line positions just to keep the operation afloat. Finally, the financial cost of overtime was becoming ludicrous.

The creative management team came up with the idea of a perfect attendance program. They knew that management mandates would do little, since the employees were not doing anything that was counter to hospital policy when they called in sick. A game, however, might stimulate the type of change they needed. At least that was their hope.

Thus, they began by talking with employees about what types of things might generate interest, such as free movie passes, lottery tickets, and gift certificates to local restaurants. Then the managers announced—with great fanfare—the opportunity for employees to win such prizes. All that an employee needed to do was not call in sick. Of course, managers were careful to tell employees that no one should come to work when ill, but that the program was intended to encourage people not to take unscheduled days off for nonmedical reasons. And the managers put money behind the program. In the first year, they spent almost $2,000 publicizing the program and investing in prizes.

Did it work? By all accounts, the program was a huge success. Employees enjoyed the competitive aspect of the game, as well as the generous prizes. Financially, the program resulted in a decrease of more than $11,000 in sick pay and a decrease of almost $77,000 in overtime pay. Perhaps of greater value, the results soon became a source of pride for employees and the management team. This, in turn, led to a change in work habits that increased overall morale—and productivity.

In either case, the important thing is to recognize employees' efforts in this area publicly and to send an explicit message that positive behavior is rewarded. The key to the success of perfect attendance initiatives is the communication of the goals and the rewards. While the program may be perceived as a game or an amusing distraction from the routine work environment, it offers management the opportunity to introduce positive change while recognizing employees' importance to the overall mission of the operation. This chapter's best practice underscores the value of such a program.

Worker Safety

Worker safety programs were originally introduced to minimize injuries on the job. The financial benefit was the projected savings in sick pay and the resulting lower workers' compensation rates. Such programs also produced a surprisingly larger benefit, however, in the form of increased productivity. These gains were realized not only from the reduced number of injury-related absences, but also from the focus on working more intelligently (and safely).

Worker safety programs, like perfect attendance programs, are designed as games in which employees compete. The game's (and management's) goal is an injury-free workplace. Thus, achievement is measured in terms of consecutive days with no staff injuries. Of course, as in promoting perfect attendance, employees are instructed not to hide injuries for the sake of winning.

The medium for communicating as well as tracking progress is usually a large sign in the kitchen that is updated daily with the number of days since the last injury. The threshold for rewarding employees varies inversely with the severity of the problem within the department but generally follows the same pattern as perfect attendance programs; the goal is to achieve a perfect score for one, three, or six months.

What do employees get? Again, it varies based on the size of the operation and the potential savings, but prizes usually come in the form of a foodservice-department pizza party or individual gifts such as lottery tickets or movie passes. In cases where the achievement is particularly notable, such as going for a full year without any injuries, management may reward employees more generously in the form of gifts or even a raffle for a large-ticket item such as a big-screen television.

Recognition Programs

Perfect attendance and worker safety programs are successful because they recognize employee effort, which ultimately contributes to overall behavior modification within the operation. Not all recognition programs need to be this complex. Any practice that recognizes positive behavior and communicates to everyone that such behavior is rewarded can lead to productivity gains.

The anchor for such programs can apply to any area of the operation. In the cafeteria, for example, customer satisfaction may be a useful measure. Using customer feedback cards or third-party surveys, management can gain important information regarding service and those who successfully deliver it. It is then up to management to determine how to recognize those who are making positive impacts on customers.

In the back of the house, programs can be designed to encourage increased productivity in almost any facet of the kitchen operation. On the dish machine, for example, employees might compete on how quickly (and correctly) they can manage the throughput of dishes. In the stockroom, recognition might take the form of rewards for the best-organized shift.

Finally, there is a type of recognition program that seeks to recognize everyone simply as being a member of the foodservice operation. Sometimes these programs take the form of pins given out on National Foodservice Worker Day or perhaps when a renovation project is completed. While not specifically targeted at a certain behavior, such programs are useful in building pride and a sense of identity within the department. The need here is to apply such programs discriminately, lest they be perceived as lacking meaning.

Clearly, programs geared toward recognizing employees vary widely. The only limitation is management's creativity. The key is to make the linkage between desirable behavior and a positive response clear. This is a tried and true formula. And as most experienced managers already know, companies that use some type of productivity-enhancement program, including recognition programs, have been shown to be far more productive than those that do not.[3]

CHAPTER SUMMARY

Productivity can be measured most accurately through the ratio of selected outputs and inputs. The selection of output and input variables affects the utility of the productivity statistic. Moreover, productivity measures that include only one output and one input variable may be useful in certain applications and can be applied in intertemporal analyses but may be extremely limiting in other ways.

A variety of conventions have been used in on-site foodservice to simplify the calculation of productivity. In attempting to keep the calculation simple, however, operators are too often faced with issues that lead to inaccurate determinations. Even when attempting to define something as seemingly basic as meals per labor hour, quantifying precisely what a meal is can lead to a variety of problems, particularly for certain segments of on-site foodservice.

A robust measure of productivity that alleviates the majority of problems and inaccuracies of other productivity indices includes period-specific revenue, food cost, labor cost, related operating expenses, apportioned profit or management fees, and apportioned investment. This aggregate, total-factor productivity statistic allows managers to make comparisons intertemporally as well as among multiple units without subjective justifications.

A number of programs can be introduced to increase productivity in the on-site workplace. Perfect attendance and worker safety initiatives are good examples of such programs that produce positive results. Other programs, some of which are intended to recognize employees for achievements in specific areas and others that are more global in reach, also increase productivity.

KEY TERMS

productivity
partial-factor statistic
multiple partial-factor
    statistic
sales per labor hour

meals per labor hour
equivalent meal value
floating equivalent-meal-
    price factor

intertemporal analysis
amortized leasehold
    improvements

REVIEW AND DISCUSSION QUESTIONS

1. How might productivity be measured in a manufacturing plant? Why is this more difficult in foodservice?

2. When might sales per labor hour be a useful statistic?

3. Describe a partial-factor productivity measure, different from those identified in the chapter, that might have utility for a specific aspect of managerial analysis. What are the limitations of this measure?

4. Why is counting meals so problematic in certain segments of on-site foodservice? In which segment is this determination easiest?

5. How should an accurate EMV be determined? Should it be different for different on-site operations?

6. How can intertemporal analysis be most useful for an on-site foodservice manager? Give several examples.

7. Given the robustness of the aggregate productivity measure described in the chapter, why don't more managers use it?

8. What type of operation might use a perfect attendance program? What type of operation would benefit most?

9. Can multiple productivity enhancement programs be run concurrently? If not, which should have priority? If so, which ones are most important?

10. Design a recognition program specific to a back-of-the-house area that would likely lead to improvements in productivity. Be sure to discuss your program's implementation strategy, the reward for employees, and an estimate of both the program's cost and benefits.

END NOTES

[1] Portions of this chapter were adapted from Reynolds, D. (1998). Productivity Analysis. *Cornell Hotel and Restaurant Administration Quarterly, 39*(3), 22–31.

[2] For a conceptual discussion of productivity, see Kendrick, J. (1984). *Improving company productivity.* Baltimore: Johns Hopkins University Press; Glaser, J. L. (1993). Multifactor productivity in the utility services industry. *Monthly Labor Review, 116*(5), 35–48; and Kozicki, S. (1997). The productivity growth slowdown: Diverging trends in the manufacturing and service sectors. *Economic Review, 82*(1), 31–46.

[3] Guzzo, R. A., Jette, R. D., & Katzell, R. A. (1985). The effects of psychology based intervention programs on worker productivity: A meta-analysis. *Personnel Psychology, 38*(2), 275–291.

CHAPTER 10

LEADERSHIP AND MOTIVATION

While the previous two chapters detailed many of the most critical issues relevant to managing people, one topic that has been omitted purposely is that of leading and motivating people. Any savvy manager knows that even with the best human-resource planning, recruitment, selection, and training and development programs in place, and even with sound productivity standards and programs fully implemented, a staff will simply not produce desirable results without the necessary leadership and motivation. With the labor challenges facing the entire foodservice industry today, this is more critical than ever to any on-site foodservice operation's success.

This chapter first delineates the often fuzzy difference between leadership and management. While entire textbooks cover the topic of leadership, this chapter addresses some of the challenges that are most endemic in on-site foodservice and describes some useful theories.[1] Next, it examines contemporary motivation theories in depth. The intent of this approach is to offer managers a better understanding of how motivation can be viewed. The related topics of power and influence are discussed, again within the framework of leading and motivating today's diverse foodservice employees. A best practice is included as an illustrative example.

The notion of influence is then extended to the practice of using supervisors' expectations to motivate subordinates. Intriguingly, managers can harness this phenomenon, called the "Pygmalion effect," simply by effectively expressing their positive expectations regarding their employees' work-related behavior. Finally, the discussion—underscored by another best practice—segues into the related subject of empowerment and provides examples of how foodservice managers can best embrace empowerment in the workplace.

LEADERSHIP AND MANAGEMENT

Leadership here refers to the *process of guiding and directing the behavior of people* in the foodservice department. More specifically, leadership produces useful change in the department that permeates the host organization, improving the experience of the customer, the employee, or anyone else involved in the system; this may include vendors or other department heads. In a healthcare setting, it may extend even to friends and relatives of patients who aren't organizationally involved in the healthcare-service chain. On a more macro level, this change might affect communication within and between departments or transcend the way a customer or patient is treated.

Good management integrates quality leadership into business processes. Following the traditional functional model, management involves (1) **planning,** (2) **organizing,** (3) **directing,** and (4) **controlling.**[2] For example, budgeting involves establishing a plan for the upcoming year. Staffing the department to meet the needs of the organization is a function of the organizing process. Ensuring that the goals and objectives of the department are met, often through problem solving, is part of both directing and controlling. In sum, these functions facilitate the utilization of resources to achieve the goals of the department—and the organization.

In muted contrast, leadership involves setting a direction for the foodservice department—one that corresponds to the direction of the organization, aligning everyone in the department with that direction through effective communication, and motivating people to action. Hence, all foodservice managers need to play a leadership role in order to be effective. Generally, managers who are not effective leaders may be successful in the short term but cannot sustain quality delivery of service and food consistently over time. These individuals—commonly referred to as **flash-in-the-pan managers**—often simply do not understand the need for effective leadership in on-site foodservice.

Indeed, on-site foodservice management once followed the same model as the old style of foodservice management in more traditional segments. That is, managers often adopted the "my way or the highway" approach. In the early years, this approach shunned leadership and empowerment of any type. Managers were dictatorial and treated employees with denigration and distrust. The problem is that in today's flatter, leaner operating environment, a manager cannot make all the decisions and rule over every facet of the department, at least not if the ultimate goal is to provide a quality experience for customers.

In cutting-edge foodservice departments, the line differentiating manager and leader is often unseen by employees. Furthermore, leadership may not necessarily be limited to the foodservice director: Supervisors have leadership positions, and in some cases line employees serve in a variety of leadership roles. The result is typically a more cohesive work environment that espouses the goals of the organization and enjoys the benefits of a team mentality such as increased employee retention, greater job satisfaction, and enhanced productivity.

In the past several years, more and more seasoned managers and theorists have suggested that the move from management to leadership is necessary in order to create and maintain a competitive advantage, particularly in an industry such as foodservice where it is relatively easy to replicate what others are doing. Indeed, it is the *service* rather than the food that is most affected by leadership and that most dictates success. This is arguably truer in the on-site segment than in any other area of the global hospitality industry. There are a host of perspectives on methods of discerning management and leadership. One of the more straightforward and applicable approaches is depicted in Table 10.1. These are not scientifically derived differences, but they do help in considering how a manager might become more leadership-oriented. These distinctions offer particular utility for managers who have risen through the foodservice ranks and have honed their management skills through a combination of trial and error, intuition, role modeling, and education. Self-reflection on how managerial characteristics differ from leadership characteristics can spark behavioral

TABLE 10.1

Some Characteristics of Managers versus Leaders

Manager	Leader
Administrates	Innovates
Focuses on systems and structures	Focuses on people
Relies on control	Inspires trust
Possesses short-range view	Maintains a long-range perspective
Asks how and when	Asks what and why
Watches the bottom line	Watches the horizon
Imitates	Originates
Accepts the status quo	Challenges the status quo
Does things right	Does the right thing

SOURCE: Adapted from Bennis, W. G. (1989). Managing the dream: Leadership in the 21st century. *Journal of Organizational Change Management 2* (1), 7.

changes in any manager's quest to move toward a larger leadership role in the foodservice operation.

LEADERSHIP THEORIES

A handful of leadership theories apply directly to the unique world of foodservice management. More to the point, a broad understanding of specific leadership theories can be extremely useful in appreciating why some approaches work better in the multifaceted operations of any foodservice department. Three types of leadership theories are discussed: trait theories, behavioral theories, and situational theories. Examples of each type are presented.

Trait Approaches to Leadership

The first set of theories relate directly to the discussion of the differences between managers and leaders by focusing on a leader's personality and attributes. The goal of these theories, referred to as "trait theories," is to identify traits or attributes of leaders regardless of the situation or circumstances in which they are deemed effective. These theories include transformational leadership and charismatic leadership.

Transformational leadership theory suggests that leaders engage in two general sets of activities. One set includes **transactional leadership,** the day-to-day leadership activities of addressing daily tasks and motivating followers accordingly. These activities are considered transactional because they focus on an exchange or transaction between the leader and followers. Followers perform the tasks in exchange for the direction, resources, and rewards the leader provides.

This exchange involves the leader in providing **contingency rewards**—rewards based on or contingent on task performance. These contingency rewards are recognized by all parties to reward positive behavior and discourage inappropriate actions. In such a market-exchange environment, compliance is the norm and failure to follow the rules generally results in systematic negative results such as progressive discipline procedures.

A common application of transactional leadership is the practice of **management by exception** (MBE). Leaders who employ MBE interact with employees and intervene in their work only when things go wrong. In such a setting, there is little goal setting or positive reinforcement; rather, the leader relies almost exclusively on discipline and punitive approaches.

Transactional leadership has its place, but its application must be appropriate to be advisable. Many foodservice managers have long used such an approach, since transactional relationships can lead to high performance, particularly when the focus is on short-term operational outcomes. On the downside, transactional relationships do not inspire employees to exceed customer expectations through innovation and extra effort. The approach does not often inspire or energize and rarely serves to build allegiance or a team orientation. Moreover, transactional leadership is ineffective in times of change because the rules and contingencies are unclear.[3] Hence, in the current era of commonplace organizational change in both the foodservice setting and the host organization—be it healthcare, B&I, prisons, or anything else—transactional leadership is not necessarily the most advisable option.

The second set of activities is oriented toward planning and implementing major challenges in organizations. To this end, individuals who engage in **transformational leadership** inspire and excite followers to achieve high levels of performance, relying on their personal attributes instead of their official position to manage employees. Hence, transformational leaders avoid a market-exchange setting, whereby rewards are exchanged for outcomes in the transactional model, and instead rely on charisma, intellectual stimulation, and individual consideration to inspire and motivate the troops.[4]

Today, much attention is focused on managers who can bring about rapid change within a department while at the same time functioning within an environment typified by massive organizational transformations. To this end, a transformational leadership approach can be very effective. Keep in mind, however, that this trait theory is not oriented toward specific situations and rests on the assumption that the leader possesses the necessary attributes to make it work. While there is some evidence that leaders may learn transformational leadership and wield it to inspire employees to perform beyond expectations, it is still largely considered artful in its application. Similarly, as shown in Figure 10.1, many do not understand its utility. Yet, as demonstrated in this chapter's first best practice, it can be embraced to produce some impressive results.

Behavioral Approaches to Leadership

As early as the 1930s, researchers began to theorize that leadership was more a function of behavior than the result of inborn traits, recommending a keen focus on what leaders do rather than on who the leaders are. The early results of this focus classified

FIGURE 10.1 What Is Charisma? Reprinted with special permission of King Features Syndicate.

TRANSFORMATIONAL LEADERSHIP IN FOODSERVICE MANAGEMENT AT FRANKLIN MEMORIAL HOSPITAL (FARMINGTON, MAINE)

At one time, foodservice management at Franklin Memorial Hospital, a small community hospital with 70 beds, vividly exemplified transactional leadership. The foodservice manager exercised his authority as though wielding a sword. There was little discussion about how duties should be performed.

Rules governing behavior in almost every instance were very clear—it was a tight ship, as the employees often heard. This is not a criticism; the model was an extension of the traditional old-style restaurant culture described at the beginning of this chapter. On its own terms, it accomplished the department's primary duties.

The problem, however, was the growingly stark contrast developing between the foodservice management approach and ongoing changes within the host organization. The administration was knowingly taking a larger role within the community. The healthcare provider was assuming a leadership position not only in providing a broad range of medical services, but also in responding to the specific needs of its local market. Franklin devoted considerable resources to developing an impressive family birthing unit as well as to expanding its medical library, which it opened to the public for research and information. Most notably, the hospital was clearly positioning itself as both a vital member of the community and the local employer of choice.

The time for change was at hand; the entire organization needed to become flexible and progressive to be a good corporate citizen. As a result, changes were made in the foodservice department that included a new management team. The new regime gave employees an equal share in the success of the department, acknowledging that everyone would need to buy in to their vision if the department was going to be a key contributor to the hospital's emerging role in the community.

This change in the foodservice management approach was particularly fortuitous since other market forces would soon demand more sweeping changes. In recent years, the Farmington community had become home to some wonderful restaurants. Despite the somewhat isolated nature of the city, the culinary offerings were quite impressive. As a result, the hospital's employees began expecting similarly fine offerings from the foodservice staff. And the employees were not shy about voicing their demand for culinary excellence.

Thus, the foodservice management team took the problem to the foodservice employees. "How can we improve?" they asked. By all accounts, the response was amazing. Perhaps it was because the employees had been quiet for so long that they couldn't hold back. Or maybe the employees knew that the managers were now eager to effect the type of change of which they could all be proud.

The transformation that resulted not only met the administration's expectations for constructive change, but also effectively repositioned the employee cafeteria as a restaurant of choice for those inside and outside the organization. They hired an executive chef, a first for the institution. Thanks to the similarly transformational management style of the newest member of the team, the changes gained momentum with staggering effects. The entire menu was redesigned and featured healthy alternatives on a daily basis. Service delivery was reengineered, resulting in quicker service in the cafeteria. And an herb garden was planted, using a plot of land conveniently located near the loading dock.

Have the changes continued? The transformational style appears still to be working, and the evolution is apparent in both the physical plant and the service delivery. According to the administration, employees, patients, and outside customers continue to rave about the food. And the pride is palpable. Throughout the foodservice department, employees know they are the fuel behind a culinary revolution. And they don't expect the changes to stop anytime soon.

leaders into three basic leadership styles. Behavioral theories later examined the behavioral dimensions of all leaders. These theories are important to foodservice organizations because they can provide direction on training and developing current managers to become more effective leaders rather than simply selecting new ones, as the trait approaches might suggest.

Kurt Lewin and his colleagues identified three basic leadership styles: autocratic, democratic, and laissez-faire. Lewin observed that leaders use one of these basic styles when approaching a group of employees in a leadership situation. He theorized that the situation itself is unimportant because the leader's style is linked to a universal behavioral pattern regardless of the situation.

The **autocratic style** is directive, with the manager controlling the group process through unilateral decision making. Leaders with an autocratic style employ a militaristic approach in which rules and regulations define the work environment. Followers are expected to obey without questioning the policies or the outcomes. In a more participative vein, a leader with a **democratic style** takes a collaborative stance, interacting more with employees and placing less emphasis on rules and policies. Employees recognize that the ultimate authority rests with the leader, but they are encouraged to make suggestions, question current practices, and contribute to enhancing the outcomes. At the other end of the spectrum from the autocratic leader, the leader with a **laissez-faire style** leads through what might be considered nonleadership; the leader avoids power and responsibility and assumes a hands-off posture.

More recent research has applied somewhat different terminology to describe the three leadership styles. For example, an autocratic style has been labeled "boss-centered," "authoritarian," and even "dictatorial." A democratic style is often referred to as "participative" or "subordinate-centered." The laissez-faire approach is sometimes translated and referred to as a "free-rein style."

Most foodservice managers can reflect on bosses from their past who have demonstrated one or more of these styles. Here we must consider whether one style may be more effective than another in a foodservice environment. For decades, any hotel general manager worth his or her salt most likely exhibited an autocratic style. Such an approach, when not taken to an extreme, can provide the necessary structure and direction for employees working in an often tumultuous workplace. Conversely, the democratic style may be more appropriate for many since there is less need for direct supervision. The psychological freedom that results may also be more advantageous in an industry that requires service delivery to be prompt, courteous, and efficient.

Another way to think about leadership behavior is to separate those behaviors that relate to the task from those that relate to taking care of people. **Initiating structure** is the term used to describe leader behavior aimed at accomplishing task-related actions. Leader behavior oriented toward relationships is called **consideration.**[5]

These two types of leader behavior are independent of each other. As such, a leader may exhibit substantial behavioral tendencies that include initiating structure as well as consideration. Similarly, he or she may be low on one or both. Examples of initiating structure can include leaders communicating what is expected, scheduling their work tasks to meet these expectations, and assigning specific tasks to individuals to ensure that uniform procedures are followed. A leader who depicts consideration-oriented behaviors treats employees as his or her equal; the leader is friendly, approachable, and endeavors to create a pleasant and familial work environment.

This approach to understanding leadership, by isolating broad dimensions, is potentially useful for on-site foodservice management because it entails objective behaviors that managers can adopt and easily put into practice. Furthermore, this is a useful method since it suggests that managers can be trained to become more effective leaders. Finally, this two-factor approach can help managers to classify their own behaviors in order to maximize their effectiveness in the workplace.

A related and useful behaviorally based approach to understanding leadership in on-site foodservice is the **Managerial Grid,** developed by Robert Blake and Jane Mouton. The researchers dichotomized a manager's attitudes and related behaviors on the basis of two dimensions. In the classification scheme, *concern for production* involves an emphasis on output (such as menu items), cost effectiveness, and profit. *Concern for people* involves promoting interpersonal relationships, helping employees integrate in the workplace, and paying attention to issues of importance to employees.[6]

As Figure 10.2 shows, the Managerial Grid theory depicts five key styles of leadership wherein the balance of the two attitudinal categories is different. The **country club manager** focuses almost exclusively on the needs of the people and the interpersonal relationships within the workplace but devotes little time and energy

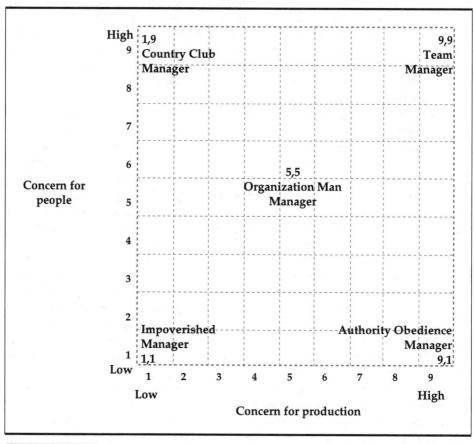

FIGURE 10.2 The Management Grid. Adapted from "The Leadership Grid" figure from Blake, R. R., & McCanse, A. A. (1991). *Leadership Dilemmas—Grid Solutions.* Houston, TX: Gulf Publishing Company, p. 29. Copyright © 1991 by Scientific Methods, Inc. Reproduced by permission of the owners.

to the quality or quantity of departmental output. In sharp contrast, the **authority-obedience manager** emphasizes production and efficiency, in large part by minimizing the need for interpersonal contact. Behaviors associated with this leadership style are similar to those of the autocratic manager discussed earlier. With behavior similar to that of the laissez-faire style of leadership, the **impoverished manager** largely abdicates responsibility for both employees and food-related products or service. The **organization man manager** strives for adequate departmental performance with modest morale levels; this type of manager does little to affect the status quo and is resistant to change. Finally, the **team manager** emphasizes high levels of quality output while concurrently focusing on developing employees as well as fostering individuals' interconnectedness.

Supporters of this leadership theory contend that the team manager is universally the most effective. It has also been suggested that experienced managers, regardless of the industry or service environment, prefer this team style of leadership. The preference for this style, however, does not mean that it is effective in a foodservice setting. It may be that situational factors alter the balance between an overarching focus on either people or production, which is supported by empirical evidence that the team-manager approach is not the most effective form of leadership in all situations.[7] Nonetheless, the Managerial Grid serves as a good reference for managers who are striving to understand what type of leadership may best suit their specific on-site foodservice operations.

Behavioral Approaches to Leadership Across Cultures   National culture can greatly influence the perceptions, expectations, and effectiveness of leadership behaviors as well as the applicability of the theories that describe them. One behavior-based theory of leadership that both facilitates an understanding of these differences and aids in the consideration of different leadership behaviors in on-site foodservice operations worldwide is the **performance-maintenance (PM) leadership theory.** Stemming from studies begun in Japan shortly after World War II, PM researchers have investigated whether leadership approaches that are effective in the United States can be applied to Japanese organizations.

To this end, researchers separated leadership behavior into two categories, P-oriented behavior and M-oriented behavior. **P-oriented behavior** is similar to the initiating structure described earlier. Its characteristics include behaviors that emphasize fast work, quality, quantity, and accuracy of output, as well as a high degree of rule adherence. Similarly, **M-oriented** and consideration behaviors are much alike. M-oriented leadership emphasizes subordinates' feelings, comfort in the work environment, stress reduction for employees, and appreciation for everyone involved. Just as with the other theories that incorporate two primary effects, such as that depicted in the Managerial Grid, both behaviors can be present; it is the greater orientation toward performance or maintenance that is of interest.

In order to understand fully the theoretical underpinnings, it is helpful to visualize the PM concepts overlaid with Lewin's autocratic, democratic, and laissez-faire styles discussed earlier. Autocratic leadership is similar to predominantly P-oriented behavior, whereas democratic leadership is more akin to M-oriented behavior. Laissez-fair leadership has a rough similarity to combined PM behavior.

Research involving PM theory indicates that the leadership behavior of high-producing groups involving lower- and middle-level managers was most often M-oriented. Interestingly, P-type leadership was even less successful in field experiments than it was when tested in actual work settings. Moreover, the findings suggest that this autocratic style was less effective in Japanese companies than in some U.S. companies.

The message of PM theory and related research is clear: National culture does influence the effectiveness of leadership behavior. This is important not only for expatriate managers working in on-site foodservice, but also for managers who have a culturally diverse workforce. As discussed in Chapter 8, foodservice in all segments is seeing greater diversity in the workplace. Hence, an understanding of how subordinates may relate to different leadership approaches on the basis of their cultural orientation is more important than ever.

On-site foodservice managers may help themselves by exploring their leadership behavior, thinking about the way they and their immediate supervisors lead. Managers commonly develop behaviors that emulate those of their supervisors. For this reason, an understanding of how one's boss leads can be quite useful (see Figure 10.3).

Situational Approaches to Leadership

Trait and behavioral approaches to leadership are helpful in understanding some of the more general aspects of leadership. As most experienced foodservice managers know, however, leadership is often dictated, or at the very least affected, by the situation in which it is exercised. In response, a number of theories have emerged that can be considered situational approaches to leadership. These include leader-member exchange, Fielder's Contingency Theory, and the Situational Leadership Model. While others exist, these three offer a number of attributes that make them particularly appropriate to foodservice-management settings.

Leader-member exchange (LMX), initially called "vertical dyad linkage," contends that leaders do not treat all subordinates alike. The differences are related to relationships between the subordinate and the supervisor that shape the behavior of both parties. For example, a subordinate, say a counterperson, who demonstrates commitment to the operation through such behaviors as good attendance, a positive attitude, and general service orientation will likely be rewarded with more of the leader's positional resources (such as confidence, concern, additional training, and possibly even more opportunities for advancement) than others who do not display such behaviors.

As these relationships develop, the leader has a propensity to develop an in-group of subordinates as well as an out-group. The in-group consists of those subordinates who have established a history of positive exchanges with the leader and in whom the leader has greater confidence. Research related to LMX suggests that in-group subordinates perceive the leader as possessing greater trust in them and report fewer challenges in the hierarchical relationship than out-group subordinates. Also, leaders extend greater consideration, in the context described earlier, to in-group subordinates. As a result, these subordinates typically assume greater responsibility in the operation, contribute more to its goals, and receive higher performance ratings than those having low-quality relationships with the leader.[8]

One of the simpler tools available for this exercise is the PM Leadership Evaluation, which is the set of questions often used in researching PM leadership theory. Answers to these questions produce two scores, one representing the respondent's P-orientation and the other representing the respondent's M-orientation. Take a few minutes to answer the questions below. Try it once to rate your supervisor's leadership style. Then ask your subordinates to answer the questions about you anonymously. Your findings may surprise you.

Leadership Evaluation

		Not at All					Very Much	
1.	Is your superior strict about observing regulations?	1	2	3	4	5	6	7
2.	To what extent does your superior give you instructions and orders?	1	2	3	4	5	6	7
3.	Is your superior strict about the amount of work you do?	1	2	3	4	5	6	7
4.	Does your superior urge you to complete your work by the time he or she has specified?	1	2	3	4	5	6	7
5.	Does your superior try to make you work to your maximum capacity?	1	2	3	4	5	6	7
6.	When you do an inadequate job, does your superior focus on the inadequate way the job was done instead of on your personality?	1	2	3	4	5	6	7
7.	Does your superior ask you for reports about the progress of your work?	1	2	3	4	5	6	7
8.	Does your superior work out precise plans for goal achievement each month?	1	2	3	4	5	6	7
9.	Can you talk freely with your superior about your work?	1	2	3	4	5	6	7
10.	Generally, does your superior support you?	1	2	3	4	5	6	7
11.	Is your superior concerned about your personal problems?	1	2	3	4	5	6	7
12.	Do you think your superior trusts you?	1	2	3	4	5	6	7
13.	Does your superior give you recognition when you do your job well?	1	2	3	4	5	6	7
14.	When a problem arises in your workplace, does your superior ask your opinion about how to solve it?	1	2	3	4	5	6	7
15.	Is your superior concerned about your future benefits like promotions and pay raises?	1	2	3	4	5	6	7
16.	Does your superior treat you fairly?	1	2	3	4	5	6	7

Add your answers to Questions 1 through 8. This number indicates your supervisor's performance orientation.

Add your answers to Questions 9 through 16. This number indicates your supervisor's maintenance orientation.

P-orientation = _____   M-orientation = _____

A score above 40 is considered high, and a score below 20 is considered low.

FIGURE 10.3   On-Site Exercise. Adapted from The performance-maintenance (PM) theory of leadership: Review of a Japanese research program by J. Misumi and M. F. Peterson, published in *Administrative Science Quarterly*, 1985 (volume 30, issue 207). Reprinted by permission of the *Administrative Science Quarterly*.

This exchange framework is valuable in that it emphasizes how leaders may treat subordinates differently and underscores the effect such behaviors may produce. Such dyadic relationships are particularly apparent in foodservice operations that have only a handful of employees. In these cases, there are typically one or two individuals who, while at the same rank as the others, enjoy a different relationship with the manager. While little research has been conducted in this setting, its conceptual framework may help foodservice managers think more about the situational factors involved in managing both large and small operations.

Integrating the importance of interpersonal exchange, Fred Fielder developed another situation-based leadership theory in the 1960s that has earned considerable respect over the past several decades. The basic premise of **Fielder's Contingency Theory** is that the fit between the leader's style and the favorableness of the situation dictates the effectiveness of workplace output. The theory separates leaders' styles into two groups, one that is task-oriented and one that is relationship-oriented, with the difference based on the leaders' primary need for gratification.

The situational favorableness consists of three dimensions. The first is the *leader–member relationship,* which is similar to the relational characteristics described in the previous discussion of LMX theory. This is the most critical variable in determining the situation's favorableness. The second is the *degree of task structure,* which refers to the number and clarity of rules, regulations, and procedures involved in accomplishing the operation's primary task. Finally, the *leader's position power,* obtained through formal authority, pertains to the leader's legitimacy in evaluating and rewarding performance and punishing noncompliance.

Situations are considered favorable when good leader–member relationships exist, tasks are highly structured, and the leader is in a strong position of power. Conversely, an unfavorable leadership situation exists when relationships are strained, little task structure exists, and the leader has little formal power. Between these two extremes, the leadership situation offers varying degrees of moderate favorableness.

Fielder's Contingency Theory suggests that the task-oriented leader has a higher likelihood of success in both very favorable and very unfavorable situations. The explanation is that in highly favorable situations the leader can easily move the operation toward the desired objective since the employees are trusting, the task is clearly delineated, and the leader's power is understood by everyone involved. In highly unfavorable situations, a leader with a task orientation is believed to be more advantageous than one with a relationship orientation, because the former will likely take charge and make decisions without soliciting input, whereas the latter will place more emphasis on making people happy but will have greater difficulty accomplishing the respective task.

For foodservice managers, this theory offers an interesting perspective. In situations where a new manager, say one with a relationship-oriented leadership style, takes the reins in an operation, an evaluation of the situational favorableness may suggest a course of action that is different from what the individual is accustomed to. Without question, an understanding of the situation in concert with an awareness of one's tendencies toward a task or relationship orientation can certainly increase the likelihood of success in terms of leading the operation to a profitable outcome.

The final situational leadership theory is depicted by the **Situational Leadership Model.** Developed by Paul Hersey and Kenneth Blanchard, the model focuses on the readiness of the subordinates as the primary conditional situation. Similar to Fielder's theory, the model employs two dimensions of leader behavior: concern with tasks and concern with relationships. On the basis of these behavioral dimensions, the model (as shown in Figure 10.4) suggests different leadership styles that are appropriate for four different levels of subordinate maturity. Followers' readiness is determined by their willingness and ability to accept responsibility related to the tasks involved.

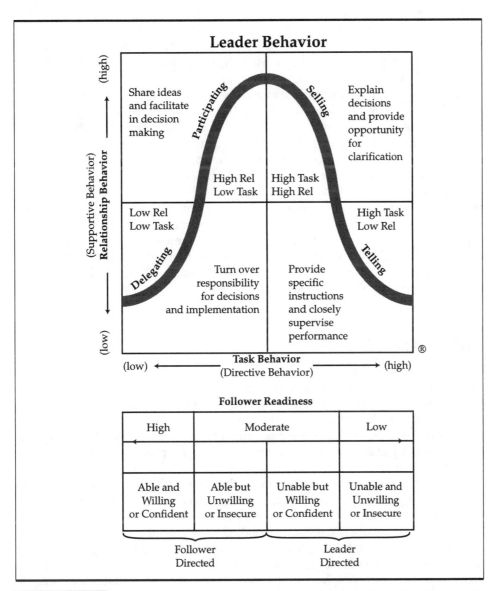

FIGURE 10.4 The Situational Leadership Model. © Copyrighted material. Adapted with permission of Center for Leadership Studies, Escondido, CA 92025. All Rights Reserved.

This might have a ring of familiarity for many foodservice managers. According to the theoretical foundations of the model, a leader should adopt a telling style of leadership with immature followers; these are individuals who are unwilling and unable to accept responsibility for completing their work. Such a leadership style focuses almost exclusively on the task, with strong initiating structure behavior, and minimizes the importance of the superior–subordinate relationship. For employees at the second level of maturity, leaders are advised to use a selling style, one in which their concern with both the task and the relationship is high. For the able but unwilling employees, those at the third level of maturity, the leader needs to use a participating style. This translates into a high concern with relationships and a low concern with the task. Finally, the most mature followers require a delegating style of leadership. These employees accept responsibility, and as a result, the leader shows low concern with both the task and the relationship.

It is likely that many foodservice managers will have the opportunity to embrace these different styles at different times, and that the maturity of the employees may have a strong correlation with their level in the organization. Nonetheless, this framework does provide insight into the difficult situation that foodservice managers face every day—one that includes employees at various points on the maturity scale who are also more responsive to different leadership styles.

As stated earlier, this is not an exhaustive list of leadership theories. The theories presented, however, serve as a general reference for managers who are trying to understand their own approach to leadership or who are interested in developing their approach in the foodservice trenches. One thing is certain: Foodservice managers need to lead effectively in order to be successful. And for most, leadership is partly innate and partly an ability learned and honed over time. A broad understanding of these theories can facilitate this education process.

MOTIVATION

Just as the foodservice manager must lead, he or she must also motivate the troops. Genuine **motivation**—defined as *the process of arousing and sustaining goal-directed behavior*—energizes, directs, and sustains human behavior in the workplace. This is not an easy endeavor. The wide range and variety of motivation theories underscore the complexity involved in understanding employees' behaviors in the workplace. Nonetheless, a manager who cannot motivate will generally not survive in today's foodservice operations.

Yet how is this accomplished? In much the same way that the previous section used leadership theories and research to help depict what may be appropriate in foodservice management, this section describes and discusses some of the more relevant motivation theories. The goal is to aid managers in enhancing their leadership skills through an understanding of what best motivates employees. Regrettably, there is no single best method. Just as there are several ways to prepare a meal, there are many ways to motivate people. And, just as the best chefs can use different techniques to prepare a meal based on the ingredients and the audience, a great manager can embrace different motivational strategies to achieve organizational objectives.

The following section offers an overview of two content theories, Maslow's Hierarchy of Needs and Herzberg's Motivational Theory. In addition, three process-related motivation theories are considered: Goal-Setting Theory, Expectancy Theory, and Equity Theory.

Maslow's Hierarchy of Needs

In the early 1940s, Abraham Maslow developed a groundbreaking theory of motivation that assumes a **Hierarchy of Needs.** As shown in Figure 10.5, these needs range from the physiological, such as hunger, thirst and sleep, to self-actualization, wherein all the lower, intermediate, and higher needs of human beings are realized and all potential has been actualized. Maslow drew primarily on his clinical background in developing his model. He believed that people's needs can be arranged hierarchically and that once a certain level is achieved, the person can only be motivated to satisfy the next highest level. Furthermore, the theory stipulates that the higher-order needs are relatively unimportant until the lower needs are met.

Interestingly, Maslow did not intend his model to be applied to the area of management, let alone foodservice. In fact, it was not until 20 years after he originally proposed his theory that he formally considered its relevance in motivating employees. Nonetheless, the theory does pose some interesting applications for motivating on-site foodservice personnel despite the complexity involved in trying to understand human needs.

As noted, physiological needs are the most basic. Examples include hunger, thirst, and sleep. From an organizational perspective, one might equate pay to the most basic requirement. Security needs, in a similar way, can be equated to benefits employees need and expect from an employer once the most basic needs are satisfied. Satisfaction of the need for love or, again from an organizational perspective, a sense of belonging might be considered next in the hierarchical order. Esteem needs are realized after the lower needs are satisfied; this might be equated to employees' need

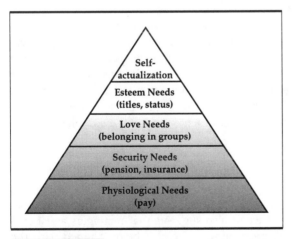

FIGURE 10.5 Maslow's Hierarchy of Needs.

for increased responsibility or status. Finally, self-actualization is what Maslow described as the culmination of satisfied lower, intermediate, and higher needs. This can be translated to the workplace in the form of personal growth and realization of one's fullest potential in the operation. Any manager would hate to lose such highly motivated employees.[9]

Herzberg's Motivational Theory

The **Herzberg Two-Factor Theory** is a simpler model of motivation than Maslow's. The theory, which was actually created as an extension of Maslow's theory, assumes that dissatisfaction and satisfaction are independent dimensions. Similarly, the aspects of the job that produce dissatisfaction are different (according to the theory) from those that produce satisfaction.

Within this theory, the satisfied employee is not necessarily a person whose dissatisfaction is nonexistent since satisfaction and dissatisfaction are evoked by different conditions. (The converse is also true.)[10] Dissatisfaction is generally attributed to extrinsic factors such as pay, working conditions, and company policies—variables referred to as **hygiene factors.** Satisfaction and motivation are derived from a different set of variables, such as attainment of recognition needs, achievement, and responsibility (classified as **motivators**). Thus, Herzberg accepts Maslow's notion of higher and lower needs but draws qualitative distinctions between the two.

From a foodservice-management perspective, Herzberg's two-factor theory arguably oversimplifies the complexities of worker motivation but also usefully draws a simpler picture of a complex landscape. For example, Herzberg suggests that job dissatisfaction might stem from company policy, supervision tactics, salary, interpersonal relations, and working conditions. Most managers have heard such reasons for an employee leaving the foodservice operation during exit interviews with subordinates. Herzberg's motivators, which he describes as achievement, recognition, the nature of the work, and responsibility, are also often cited by employees as reasons they enjoy working in the foodservice sector. The manager interested in motivating employees can embrace both categories appropriately, with positive outcomes.

Both Maslow's Hierarchy of Needs and Herzberg's Two-Factor Theory are considered content theories because they focus on the needs and, in some cases, drivers of satisfaction of employees. While somewhat limited in their scope, these theories are useful because they simplify what motivates many in the foodservice sector. The next set of theories focuses more on the cognitive antecedents that affect motivation and performance. These contemporary iterations of what originated as process theories also recognize that various factors affect one another, owing to the complexity of the human psyche.

Goal-Setting Theory

Goal-Setting Theory maintains that an employee's conscious objectives will most likely influence the resulting performance. It also suggests that managers assist employees in developing goals that are motivating while also being aligned with the common goals of the organization. While earlier theories have received only modest empirical support, practitioners as well as researchers recognize goal-setting theory as possessing substantial validity and practical usefulness.

As depicted in Figure 10.6, the theory suggests that employees' motivation begins with their values and related value judgments. These lead to emotions that, in turn, transform the values and judgments into intentions or goals. As a result, behavior is enacted that satisfies these goals. Finally, results in the form of consequences or feedback (either positive or negative) are received.

Goal setting and related approaches such as management by objective have direct application to foodservice management. However, in order to be effective, motivating goals must be clear and specific. Furthermore, they must be realized through dialogue between manager and employee; feedback regarding progress toward the goals must be frequent and clear. This requires an active role for the manager that cannot be delegated or treated casually. Moreover, goals must be realistic while representing a stretch in terms of performance. In a nutshell, goals must be tangible, measurable, and achievable.

Expectancy Theory

Initially developed by Victor Vroom in 1964, the **Expectancy Theory** of motivation makes three basic assumptions about employee behavior. First, employees perceive a relationship or **instrumentality** between a certain work behavior and some benefit or payoff that will result from the behavior. Second, there is **valence,** or expected utility, associated with this positive outcome. Third, employees subconsciously or consciously assess the degree of likelihood of the reward; Vroom calls this the **expectancy.**[11] Hence, the direction and intensity of the behavior are functions of the expectation that certain actions will lead to the goal, thereby creating an assumed value for the goal object. The theoretical relationship of the concepts of valence, instrumentality, and expectancy is illustrated in Figure 10.7.

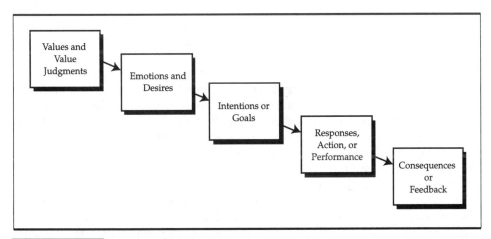

FIGURE 10.6 Goal-Setting Theory of Work Motivation.

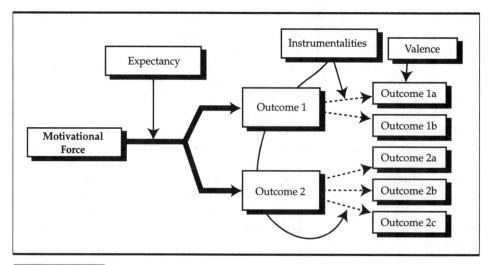

FIGURE 10.7 Expectancy Theory of Motivation.

Vroom's theory deviates from other motivation theories in that it depicts a process of cognitive inputs that reflects individual differences. In addition, it does not take a simplistic view of employee motivation. Most notably, it puts an extremely limited onus on the manager and attributes most of the cognitive function to the employee. In a foodservice operation, this translates into a need for the manager to possess an awareness of the unique effects that his or her actions may have on each employee. On a related note, each employee's valences, instrumentalities, and expectancies are different. Hence, the foodservice manager, armed with this knowledge, must strive to understand the differences and similarities among his or her employees.

Equity Theory

Taking a different approach, Stacy Adams' **Equity Theory** (sometimes referred to as "Inequity Theory") focuses on the discrepancy workers perceive between their work and the rewards or compensation they receive and their coworkers' level of work and level of rewards. When an employee perceives that he or she is working more than a coworker for the same reward or pay, the theory suggests that the employee will lessen his or her productivity as part of an automatic response mechanism. Similarly, an employee who perceives that a coworker receives better pay for the same job will consciously reduce his or her work effort. On the other side of the coin, the theory holds that employees will perceive equity and satisfaction when they see that their outcomes, proportionate to their effort, are the same as those of others. (See Figure 10.8.)

Recent research reflects what many foodservice managers have found when attempting to apply the theory to aid their motivational efforts. That is, there appear to be three different types of employees, each possessing a different sensitivity and orientation to differences in perceived equity. "Equity sensitives" are those who fit the description offered by the original theory. Lack of perceived equity for these individuals creates

Inequity

$$\frac{\text{Person's outcomes}}{\text{Person's inputs}} \;<\text{or}>\; \frac{\text{Other's outcomes}}{\text{Other's inputs}}$$

Equity

$$\frac{\text{Person's outcomes}}{\text{Person's inputs}} = \frac{\text{Other's outcomes}}{\text{Other's inputs}}$$

FIGURE 10.8 The Role of Outcomes and Inputs in Equity Theory.

considerable productivity issues. "Benevolents," who might be thought of as givers, are comfortable with a situation that gives others a higher outcome/input ratio than their own. Finally, "entitleds" are those who are comfortable with an inequitable situation so long as they are benefiting in the form of a greater outcome/input ratio.

We find a variety of each of these types in a typical on-site foodservice operation. The challenge for the foodservice manager is, first, to strive for equity among all employees and, second, to recognize employees who might possess different equity sensitivity. Through such efforts, the manager can likely use an understanding of Equity Theory to maximize staff performance and minimize issues that often arise as a result of inequity in the workplace.

In subtle ways, each foodservice employee is likely motivated somewhat differently. However, these theories can provide insight to managers attempting to maximize productivity as well as employee development and retention. As alluded to earlier in the discussion of leadership, however, theories on how to manage foodservice employees enhance an operation only if the manager uses them correctly. Additionally, standard practices of sound management, including ethical treatment of employees and conscientious use of power and influence, cannot be replaced by any management approach, whether it is theoretically viable or not.

POWER AND INFLUENCE

Foodservice managers can learn to lead and motivate their employees more effectively by understanding the effects of power and influence that are exchanged whenever a group of people is assembled. Simply stated, **power** is the *potential ability to influence someone else.* **Influence** is the *process of affecting another person's thoughts, behavior, or feelings.* Like leadership and motivation, the concepts of power and influence are varied and, at times, murky. Despite different thoughts on the subject (it would seem that many managers and most academics have their own perspective on how to define power and influence in the workplace), understanding the various types of power and the associated influence tactics can make the task of leadership in the foodservice management ranks more comprehensible.

More than 40 years ago, social psychologists John French and Bertram Raven identified five categories into which they divided sources of interpersonal power (reward, coercive, legitimate, referent, and expert). While subject to interpretation and various applications over the years, describing and analyzing these categories is still useful in understanding the sources of power a manager uses to influence subordinates.

Reward power is based on the manager's access to and ability to control the distribution of rewards that subordinates want. Quite simply, subordinates comply because they desire what the manager can give them. The most common examples are salary increases, bonuses, and promotion; a manager can allocate such things unilaterally in many cases, thereby underscoring his or her reward power. The effective use of reward power is predicated on the manager's clearly defining what behavior is rewarded and then explicitly connecting the behavior with the reward for everyone involved.

Coercive power is based on the manager's ability to punish subordinates. Employees comply simply because they are afraid of the penalty of noncompliance. Coercive power is typically linked with abuse since it generally originates with verbal threats or abuse.

Legitimate power, which loosely corresponds to the concept of authority, is based on the manager's formal position. Employees comply because of their belief in the legitimacy of the manager's role. This type of power is somewhat unique because it results from mutual agreement: The employees feel they are obliged and expected to accept this power. Along with legitimate power comes the ability to wield both reward and coercive power. Legitimate power is often accentuated by organizational symbols such as a different style of dress or preferred parking. For example, in a foodservice setting, the manager often wears professional attire while the employees don uniforms.

Referent power, the hardest of the five to explain, is based on the person's attractiveness to others. In other words, a manager has referent power because others identify with, like, or want to be like the manager. This type of power is closely linked to and forms the basis for charismatic and transformational leadership discussed earlier in this chapter. Referent power is also unique in that the manager does not always possess it. Subordinates, too, can possess and use referent power, although it has greater impact when paired with legitimate power.

The last of the five, **expert power,** is similar to referent power in that it is personal and does not depend solely on the organization. It exists when an individual has knowledge or information that another needs. Expert power is particularly important because it can be shared; the knowledge or skill can be readily spread to others. It is also the most important in enhancing the overall skill level of a group or department.

A foodservice manager who possesses expert power (where the expertise is relevant to the operation) and who is willing to share this power openly is perhaps the best and most powerful manager. He or she can use this power to develop others, which in turn enhances the quality of the food and service. When combined with legitimate and referent power, expert power is tremendously important to the advancement and evolution of an organization, making this combination a worthy goal of any foodservice manager looking to ensure the long-term success of the operation and his or her career.

Any discussion of how employees are influenced must include ample consideration of the ethical use of power. At the beginning of this section, we defined power as the potential to influence someone else. With power comes responsibility; with responsibility comes the opportunity to affect others for good or ill. Almost everyone can describe the worst boss he or she ever had. Most such descriptions portray managers who did not use power ethically.

The ability to influence others also carries an implicit responsibility to know that not everyone is influenced in the same way. For example, employees who think they have no other options for employment are particularly vulnerable to a manager's power and influence. Cultural values also have a tremendous effect on employees' susceptibility to influence. Managers must use caution when wielding their power, whatever its form.

PYGMALION IN FOODSERVICE

In Greek mythology, Pygmalion was a celibate prince of Cyprus who sculpted an ivory statue of the ideal woman. Pygmalion's sculpture was so perfect that he fell in love with his masterpiece. Seeing the prince's deep love, the goddess Aphrodite brought the statue to life, naming her Galatea, and the two lived happily ever after.

In a management context, a **Pygmalion effect** is said to occur when a person (e.g., a manager) expresses positive, but realistic, expectations toward another person or group (e.g., subordinates) that result in positive outcomes reflecting these expectations. The person expressing the positive expectations serves as the sculptor (Pygmalion) in shaping the outcome or performance of the target individual's behavior.[12]

The Pygmalion story has many applications in foodservice management. Recent research, for example, indicates that managers at all levels can be positively influenced by the expectations of their superiors.[13] In addition, it appears that basic differences among subordinates make no difference in how they respond to expectations. For example, recent research conducted in an on-site foodservice setting suggests that men and women respond with the same intensity in terms of changes in their behavior.

Not surprisingly, the Pygmalion effect has a dark side. For example, what happens if a foodservice manager tells an employee that he or she expects the subordinate to fail in performing an assigned task? Research indicates that the subordinate will likely fulfill the expectations and fail.

This is known as the **Golem effect.** It describes the undesirable changes in subordinates' performance resulting from a supervisor's expressed negative expectations. The term stems from a Jewish legend wherein a creature was created and brought to life to eradicate evil but ultimately became a monster owing to its increasingly strong, corrupting power. Another reference to the Golem (in this case spelled "Gollum") can be found in mid-twentieth-century English literature. J. R. R. Tolkien used the name for a fictional character who was corrupted by possessing a magical ring and who, in turn, sought to inflict harm on others. Consonant with how the Golem is referenced in a managerial context, Tolkien showed how Gollum's iniquitous behavior ultimately led to his ruin.[14]

The Pygmalion effect can be a useful tool for motivating employees. But the potential of the Golem effect underscores the ethical and moral challenges involved in using expectations to produce changes in employees' behavior. Nonetheless, knowledge of the Pygmalion and Golem effects is important in getting the very best out of today's foodservice employees.

EMPOWERMENT

Owing to its largely intangible, multifaceted nature, the practice of management is particularly susceptible to cure-alls, panaceas that will make any manager more effective. Most managers in the foodservice trade have heard of and perhaps even tried a number of them. Total Quality Management, discussed in the following chapter, is a good example of this. It is a good idea that has its applications. It does not, however, fit everywhere and in every situation.

Empowerment has become one of the biggest leadership-related management buzzwords of the new century. It is one of the most written-about management topics and is viewed by many organizations as the latest fix to many ills. The concept is relatively simple. It extends the notion of delegation to an extreme: Managers grant employees the authority to manage certain processes or tasks in order to give them legitimate power to improve the outcomes. In foodservice, an example of what many think of as empowerment might be a manager who "empowers" employees to give complimentary food to customers who have a legitimate complaint; the intent is to increase customer satisfaction since problems can be solved more efficiently.

Such a situational example, however, does not accurately depict the fundamental motivational process that serves as the foundation of true empowerment. Empowerment, in its truest form, focuses on the intrinsic task motivation of the employee. Stated another way, empowered employees perceive that they (1) can, through behaviors that they deem appropriate, affect organizational results, (2) feel confident in their ability to enact these behaviors, and (3) believe that their actions are meaningful. This is very different from the typical empowerment approach adopted by many foodservice companies that simply grants a single authority-related privilege to front-line employees.[15]

What does this all mean? Empowerment in foodservice can be a powerful management tool when it is realized under the three conditions described above. Failure to do so is commonly perceived by employees as just another "bright idea" that translates into more work for them. For this reason, most empowerment programs in place today ultimately *de*-motivate employees.

In the world of on-site foodservice, empowerment can be applied successfully. But it requires forethought and can be used only when the manager is actively involved in the process. Furthermore, employees must trust that the manager is not trying to pass on more duties but is truly giving power to the employees; then the real objective of empowerment can be realized. The next best-practice example illustrates this.

PROBLEM SOLVING THROUGH CREATIVE LEADERSHIP AND EMPOWERMENT AT EMERSON HOSPITAL (CONCORD, MASSACHUSETTS)

Having encountered a number of operational issues, the foodservice management team at Emerson Hospital in Concord, Massachusetts (see Figure 10.9), adopted a unique approach to problem solving involving the entire department. While the food and its delivery were generally considered of high quality by most constituencies, the foodservice director and the hospital administration hoped that through this progressive management technique they could take the operation to the next level.

In order to identify what issues deserved the most attention, the supervisory staff conducted face-to-face surveys with patients, cafeteria customers, and employees who worked in the various functional areas of the department. The problems that merited immediate attention were—somewhat surprisingly—not overwhelming but did require fixing. For example, the surveyors found that the taste of the coffee served in the patient areas did not impress the patient population. Another problem was the traffic congestion in the cafeteria during peak service times.

FIGURE 10.9 The Entrance to Emerson Hospital. Copyright Emerson Hospital.

It would have been simple for the foodservice director to suggest solutions to these issues and have them put into effect. But would such an approach have yielded the best outcome? Probably not. Often the employees engaged in an operation have the best solutions, if only they're asked. The foodservice director realized this and took a novel empowering approach. He formed teams using different members of his staff to combat the key problems, with one team per issue.

For example, the team charged with improving the quality of the coffee involved a sanitation worker, two trayline/patient-services employees, a supervisor from the cafeteria, a dietitian, and a prep cook. The team mapped out the entire process of coffee making, from putting the grounds in the filters to disposing of the leftovers. The team came up with three major solutions. Implementation was easily facilitated because fellow employees perceived the solution to be the brainchild of their coworkers, not something that management was forcing them to do. Moreover, these solutions led to other suggestions from employees not associated with the problem-solving team who realized that their input was valued.

The management staff utilized different heterogeneous teams to address the other issues. The concept was that eventually all employees would be tapped to solve problems. This negated any chance that employees would feel excluded. It also allowed each person to develop in ways only tangentially related to his or her position in the department. Of greatest importance, the approach improved the quality of the department, a tangible attribute recognized by customers, the administration, and fellow employees.

MANY WAYS TO LEAD AND MOTIVATE

As we have seen, managers can choose from among a host of methods to lead and motivate their employees. What is the best way? The answer is different for every foodservice operation. The theories described above are intended to help managers understand the different approaches that can work by showing how the actions are conceptualized. Each theory, however, is founded on different interpretations of what employees need, how they act, and what the situation may dictate.

Each manager must use this information in a way that fits his or her particular operation and setting. Managers new to an operation need to exercise power differently from the way they would after being at the same site for a few years or even a few months. Similarly, managers who are more charismatic may face different challenges from the ones faced by those who lean toward a more autocratic style. If nothing else is clear from reading this chapter, the one message should be that just as every relationship is different, so is every foodservice operation—and each benefits from a unique set of leadership and motivation approaches.

The key is to remember that managers and employees are partners in the same endeavor. A profitable operation benefits everyone, just as a less than optimal operation negatively affects everyone involved: customers, employees, and managers. The role of the manager is difficult—but, properly executed, it is every bit as delicious and satisfying as a fine meal.

CHAPTER SUMMARY

Leadership is the process of guiding and directing the behavior of people in the foodservice department. It involves setting a direction for the foodservice department—one that corresponds to the direction of the overarching organization. While management includes leadership, it traditionally involves (1) planning, (2) organizing, (3) directing, and (4) controlling. Many characteristics differentiate managers from leaders. For example, managers control complexity, while leaders introduce meaningful change. Simply put, management is a functional position within an organization, whereas leadership is the *manner* in which management is carried out; not all managers are leaders.

Several leadership theories have direct application in on-site foodservice. Transactional leadership refers to a general set of leadership activities focused on addressing daily tasks and motivating followers accordingly. This exchange requires the leader to provide contingency rewards. Transformational leadership involves a set of activities oriented toward planning and implementing major challenges in organizations. Individuals who engage in transformational leadership inspire and excite followers to high levels of performance, relying on their personal attributes instead of their official position to manage employees.

One behavioral approach to leadership categorizes leadership behavior into three groups: autocratic, democratic, and laissez-faire. Another approach divides leadership behaviors into two types: "Initiating structure" is the term used to describe leader behavior aimed at accomplishing task-related actions, and "consideration" describes leader behavior oriented toward relationships. A similar approach, stemming from research in Japan, describes these behaviors as P-oriented or M-oriented. The Managerial Grid is yet another approach to categorizing and understanding leadership behaviors; styles included are organization man manager, authority-obedience manager, country club manager, team manager, and impoverished manager. Several other theories, including LMX, Fielder's Contingency Theory, and the Situational Leadership Model, suggest that a leader should adjust his or her behavior to the situation.

Understanding how to motivate employees is essential to understanding how to lead them. Maslow's Hierarchy of Needs, Herzberg's Two-Factor Theory, Goal-Setting Theory, Expectancy Theory, and Equity Theory are some of the useful motivational approaches for foodservice managers in the on-site sector. Similarly, power, defined as the potential ability to influence someone else, and influence, defined as the process of affecting another person's thoughts, behavior, or feelings, must be appreciated. There are five types of power that a manager may possess and use: reward, coercive, legitimate, referent, and expert.

Supervisory expectations that produce changes in subordinates' behavior demonstrate the Pygmalion effect, a powerful by-product of managers' power and influence. Positive expectations of employees' work-related behavior, expressed by the manager, can have substantial positive effects on subordinates' performance. Negative expectations can be just as powerful, though, and must therefore be voiced judiciously.

Empowerment, now a common buzzword, can be a very effective approach to leadership in foodservice. However, true empowerment means more than delegating authority. In order to be effective, empowered employees must perceive that their behaviors affect organizational results, feel confident in their ability to enact these behaviors, and believe that their actions are meaningful.

leadership
planning
organizing
directing
controlling
flash-in-the-pan
  manager
transformational
  leadership theory
transactional leadership
contingency rewards
management by
  exception
transformational
  leadership
autocratic style
democratic style
laissez-faire style
initiating structure
consideration

Managerial Grid
country club manager
authority-obedience
  manager
impoverished manager
organization man
  manager
team manager
performance-maintenance
  leadership theory
P-oriented behavior
M-oriented behavior
leader-member exchange
Fielder's Contingency
  Theory
Situational Leadership
  Model
motivation
Maslow's Hierarchy of
  Needs

Herzberg's Two-Factor
  Theory
hygiene factors
motivators
Goal-Setting Theory
Expectancy Theory
instrumentality
valence
expectancy
Equity Theory
power
influence
reward power
coercive power
legitimate power
referent power
expert power
Pygmalion effect
Golem effect
empowerment

REVIEW AND DISCUSSION QUESTIONS

1. Define leadership. How might this definition be altered to be more specific to foodservice management?

2. Discuss the four functions of management. What other examples would fit besides those addressed in the chapter?

3. Consider different situations in which transactional or transformational leadership approaches would be more appropriate. Which is more comfortable for you and why?

4. Considering your managerial experience, describe situations in which you have adopted autocratic, democratic, or laissez-faire styles of leadership. Were these approaches, given the situation, appropriate?

5. Using the Managerial Grid, plot yourself over time. Have you changed your grid position since you started as a manager? What has fueled these changes?

6. Describe the various leadership theories. Which has more relevance to you? Why?

7. Assume that you had to describe Maslow's Hierarchy of Needs and Herzberg's Two-Factor Theory to your staff (either past or present). Outline your presentation.

8. Describe the three process theories of motivation. Which makes the most sense (or is most useful for you)?

9. Describe each type of power. Rate yourself on a scale of 1 to 10 for each type. What might you do to increase your expert power?

10. Think about how others' expectations affect your performance. Next, gauge how powerful an effect your boss's expectations have on your workplace behavior. Does he or she verbalize these expectations? How often do you voice your positive expectations of your employees or friends?

11. Devise an empowerment program appropriate for an on-site foodservice operation, beginning with the problem and continuing through the execution of the program. Be as specific as possible.

END NOTES

[1] While the number of books on leadership and motivation is vast, some recommended texts are Steers, M., Porter, L., & Bigley, G. (1996). *Motivation and leadership at work* (6th ed.), New York: McGraw-Hill; and Yukl, G. (1998). *Leadership in organizations* (4th ed.), Upper Saddle River, NJ: Prentice Hall.

[2] While a variety of typologies exist for a manager's role, seminal work can be found in Fayol, G. *General and industrial management,* C. Storrs, Trans., London: Sir Isaac Pitman and Sons, Ltd. (original work published in 1916); and Breeze, J. D. (1985), Harvest from the archives: The search for Fayol and Carlioz, *Journal of Management, 11,* 43–54.

[3] For more on the downside of transactional leadership, see Zaleznik, A. (1990). The leadership gap. *Academy of Management Executives, 4*(1), 7–22.

[4] For more on transformational leadership, see Podsakoff, P. M., MacKensie, S. B., Moorman, R. H., & Fetter, R. (1990). Transformational leader behaviors and their effects on followers' trust in leader, satisfaction, and organizational citizenship behaviors. *Leadership Quarterly, 1,* 107–142; Bass, B. M.. (1985). *Leadership and performance beyond expectations.* New York: Free Press; and Bass, B. M. (1999). Ethics, character, and authentic transformational leadership. *Leadership Quarterly, 10*(2), 181–219.

[5] Good sources of in-depth information about this distinction include Stodgill, R. M., & Coons, A. E. (1957). *Leader behavior: Its description and measurement* (research monograph no. 88). Columbus: Bureau of Business Research, The Ohio State University; and Bowers, D. G., & Seashores, S. E. (1966). Predicting organizational effectiveness with a four-factor theory of leadership. *Administrative Science Quarterly, 11,* 238–263.

[6] The Managerial Grid is useful for a variety of applications. For other examples, see van de Vliert, E., & Kabanoff, B. (1990). Toward theory-based measures of conflict management. *Academy of Management Journal, 33*(1), 199–209.

[7] For example, see Larson, L. L., Hunt, J. G., & Osborn, R. N. (1976). The great hi-hi leader behavior myth: A lesson from Occam's razor. *Academy of Management Journal, 19,* 628–641; and Owens, J. (1981). A reappraisal of leadership theories and training. *Personnel Administrator, 26,* 78–82.

[8] A good summary of related findings can be found in Liden, R. C., Wayne, S. J., & Stilwell, D. (1993). A longitudinal study on the early development of leader–member exchanges. *Journal of Applied Psychology, 78*(4), 662–674.

[9] For more on applying Maslow's Hierarchy of Needs to the workplace, see McGregor, D. (1985). *The human side of enterprise: 25th anniversary printing.* New York: McGraw-Hill Book Company.

[10] For more on this, see Katz, D., & Kahn, R. (1966). *The social psychology of organizations.* New York: John Wiley & Sons.

[11] For more on this topic, see Vroom, V. (1964). *Work and motivation.* New York: John Wiley & Sons; and Eisenberg, E. & Goodall, H. L. (1993). *Organizational communication: Balancing creativity and constraint.* New York: St. Martin's Press.

[12] The phenomenological foundation of the Pygmalion effect is nested in Merton's theory of the self-fulfilling prophecy (see Merton, R. K. [1948]. The self-fulfilling prophecy. *Antioch Review, 8,* 193–210). For more information on the Pygmalion effect in management, see Eden, D. (1990). *Pygmalion in management:*

Productivity as a self-fulfilling prophecy. Lexington, MA: D.C. Heath and Company; and Reynolds, D. (2001). Integrating "My Fair Lady" with foodservice management: A theoretical framework and research agenda. *Journal of Foodservice Business Research* 4(4), 263–302.

[13] For example, see Reynolds, D. (2002). The good, the bad, and the ugly of incorporating "My Fair Lady" in the workplace. *Advanced Management Journal, 67*(3), 4–14.

[14] Tolkien, J. R. R. (1954). *The fellowship of the ring (Part I: The lord of the rings).* Boston: Houghton Mifflin Company.

[15] A number of sources are available that describe both the theoretical and applied underpinnings of empowerment, including Carlzon, J. (1987). *Moments of truth.* Cambridge, MA: Ballinger Publishing Company; Sparrow, R. T. (1994). Empowerment in the hospitality industry: An exploration of antecedents and outcomes. *Hospitality Research Journal* 17(3), 51–73; Spreitzer, G. M. (1995). Psychological empowerment in the workplace: Dimensions, measurement, and validation. *Academy of Management Journal* 38, 1442–1465; and Thomas, K. W., & Velthouse, B. A. (1990). Cognitive elements of empowerment: An "interpretive" model of intrinsic task motivation. *Academy of Management Review* 15(4), 666–681.

INTERNAL CUSTOMERS AND SYSTEMS

PART IV

TRENDS AND CHALLENGES FOR TODAY AND THE FUTURE

In Parts I through III, we have seen how on-site foodservice has evolved into a modern industry serving both external and internal customers. We have described and analyzed systems of on-site foodservice operation in the light of best industry practices that combine quality and operational efficiency. Now the focus turns forward in an effort to identify the trends and challenges of today that will shape the future of the industry. While some think that predicting the future is like gazing into a crystal ball, we can undoubtedly anticipate coming developments on the basis of what many operators are encountering today.

The first of these trends is a growing emphasis on quality. Stemming from a movement in the manufacturing sector begun decades ago that has slowly crept into the service industries, "quality" has become more than a buzzword. In many cases, the focus on quality is evident, from corporate mission statements to on-site operators' quality guarantees to their customers. Chapter 11 talks about this trend in detail, with best practices that show how quality is at the fore in one leading onsite operation. The chapter also describes a variety of tools that can aid in the push toward quality in all areas of foodservice.

Similarly, the recent proliferation of senior living centers is gaining momentum like a speeding locomotive, bringing with it a host of on-site foodservice operations. As the number of older citizens continues to grow, so will the need to cater to this segment. The trend here, as discussed in detail in Chapter 12, is toward specialization, with attention paid to the target market's preferences and habits.

As witnessed by many on-site foodservice operators, host organizations are looking for a portfolio of hospitality services, of which foodservice is only one part. Chapter 13 explores this trend and delves into a variety of issues in support services from a very different perspective. This is a particularly important topic for managers who are looking for new economies within their operations that will afford them opportunities to reduce costs while delivering a higher level of overall service.

Finally, this part concludes with a chapter devoted to emerging technologies. The computer applications discussed in previous chapters, while impressive by today's standards, only open the door to what is possible. Mind-boggling as the idea of a system that seamlessly links purchasing, production, and inventory management once was, the possibilities for the future in equipment and software are staggering. The chapter concludes by extending these possibilities into the more distant future, which undoubtedly holds wonderful opportunities and continuing challenges for those who wish to remain on the cutting edge of the world's most dynamic and exciting industry.

CHAPTER 11

FOCUS ON QUALITY

Perhaps the most interesting event of the 1980s for on-site foodservice operators was the introduction of quality initiatives in many host organizations. Some viewed the new programs as a fad (and in some instances, organizations adopted such initiatives too hastily). For others, however, the focus on quality offered a new way of looking at things, and also provided innovative tools to aid in the management of complex systems such as those in on-site foodservice management.

This chapter, then, follows the focus on quality from promise to implementation in foodservice management. Beginning with a definition, the discussion addresses the leading programs that organizations can use to link quality with day-to-day operations in an effort to shape outcomes more artfully. The best practices in this chapter are particularly impressive in that they illustrate what is possible, and the considerable work that is involved, in adopting a focus on quality.

After addressing the key success factors in implementing quality initiatives, the chapter concludes by reviewing a variety of tools designed for quality management. Fortunately, these tools can be used as stand-alone applications, appropriate for different facets of the operation, or they can be introduced sequentially by managers looking to leverage the potential advantages of a systemwide quality approach. In either case, the benefits of embracing quality as a vehicle for organizational improvement are readily apparent.

QUALITY DEFINED

Quality means different things in different contexts. One person might define quality as a specific attribute, such as the degree of marbling in a high-quality piece of beef. Another person might define quality in the context of a specific aspect of an employee's job performance. "That clerk is a quality employee; see how much time she spends with each customer!" In manufacturing, quality is sometimes viewed as freedom from deficiencies. Still another person might look at quality more generally, perhaps on the basis of the price–value relationship.

In on-site foodservice, quality is best defined as *achieving excellence on the basis of satisfaction, functional-status outcomes, and price.* Working within this definition, satisfaction pertains primarily to the customer but also to the foodservice provider, especially regarding a reasonable profit potential. Functional-status outcomes entail basics such as food safety. Finally, price speaks to the importance of perceived value as part of the quality equation.

While this definition is reasonable, it still integrates the potentially subjective concept of "excellence." In many ways, this makes for a moving target since many gauge excellence on the basis of expectations. This seems unfortunate since it suggests that the whole notion of quality foodservice is nebulous. While it would be nice if we could integrate a definition that was more akin to a manufacturing-attribute definition, such as "falling within a tolerance range of ±3/1,000 of an inch," we can also appreciate the continually escalating standards implied by our definition. Foodservice operators always want to differentiate themselves positively, so a somewhat subjective attribute may be a good thing.

Using this definition, then, how is quality measured? There is no one answer, but most measuring mechanisms fall into four general categories: customer-based measures, detection-based measures, financial-based measures, and process-based measures.

Customer-based measures are based on customer satisfaction. This measure is near and dear to every hospitality enterprise, since customer satisfaction is a cornerstone of success. Common examples of customer-based measures include surveys, frequency of purchases, complaint analysis, focus groups, and mystery shoppers. Most find that a combination of these is the most accurate measure of quality.

The key is to treat customer-based measures as constructive information and to use the information as an agent for change. The value of this should be readily apparent, since it will inevitably lead to increases in quality. A former president of Xerox said it best: "If you can stand the pain, look at your business through the eyes of your customer."

Detection-based methods of measuring quality focus on the tangible aspects of food and service. For example, customers in an on-site setting may be required to dine within a given amount of time, say 30 minutes. Thus, a detection-based measure would assess whether all customers receive their food with ample time to consume it, given the time constraint. Another detection-based measure might relate to food temperature. Assessment would represent a comparison of actual serving temperature versus ideal serving temperature.

The next approach to measuring quality is through **financial-based methods,** which focus on financial analyses. Relating to the customer, some foodservice organizations use this approach to measure gains in sales and, correspondingly, profits. Representative incremental increases in both categories suggest that customer loyalty and satisfaction are increasing while also serving as indicators of a financially healthy operating status.

Process-based methods of assessing quality look at systems, methods, approaches, and processes as opposed to measures more fully focused on the end product. This is a particularly useful measure because it focuses on continual assessment of quality improvement. Tools used in this method include statistical-process control charts, among others.

PROGRAMMATIC APPROACHES TO QUALITY

Measurement is important to any quality initiative, but the manner in which quality is approached is potentially even more critical. The three most common systems are Total Quality Management, Picos, and the various systems designed by the International Organization for Standardization.

Total Quality Management

Perhaps the best known, or at least the most often referred to, approach to quality is **Total Quality Management** (TQM). Dr. W. Edwards Deming planted the seeds for the development of TQM while teaching statistical quality control as part of the wartime production effort during World War II. Deming realized that managing quality meant more than just reworking products or services until they were acceptable. He fostered the notion that quality could be approached as a process and that such efforts would result in better, more efficient, and less expensive outcomes.

Attempting to define TQM is problematic, since its many different proponents often use voluminous descriptions with specific reference to the respective business applications. Nonetheless, a good, working definition of TQM is *a cooperative form of conducting business that maximizes the talents and capabilities of both line employees and managers to continually improve quality and productivity through a team-based, participative approach.*

Even this definition seems a bit long, but it does communicate the key elements of TQM. These include a committed and involved management; an unwavering focus on internal and external customers; effective involvement and utilization of the entire workforce; continuous improvement of business processes, as well as processes of the production of food and delivery of services; and, finally, integration of performance measures as a means of evaluating all processes. In sum, these steps lead to an iterative process that results in a continual evolutionary approach in which quality is the central focus.[1]

Many businesses put TQM into operation through a **continuous quality improvement** process (CQI). CQI is sometime used synonymously with TQM, but in reality it is just one of the more popular applications of a general TQM approach. As depicted in Figure 11.1, CQI is best represented as composed of four basic operations

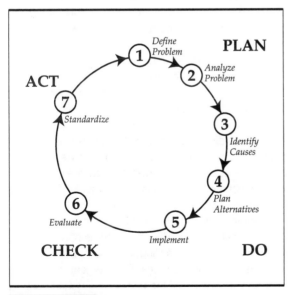

FIGURE 11.1 A Seven-Step Continuous Quality Improvement Process.

that Deming articulated as key to any quality process: plan, do, check, and act. These operations can be further broken down into seven distinct steps.[2]

Executing this CQI process is often best accomplished through **quality circles.** A quality circle is a group, typically involving six to nine people, that meets regularly to discuss problems, challenge assumptions, seek solutions, and outline steps for implementing new initiatives. Participation is strictly voluntary, but once a commitment to participate is made, members must take the role seriously. In addition, membership in a quality circle requires that all participants will assume an active role in the discussions and that everyone will strive to reach decisions through consensus. Quality circles are considered a worthy approach to problem solving even in the absence of a dedicated TQM program, but they have an even richer impact when used in conjunction with global quality initiatives. A good example is presented in this chapter's first best practice.

BEST PRACTICES IN ACTION

QUALITY AND COMMUNICATION AT OHARA MEMORIAL HOSPITAL (KYOTO, JAPAN)

Even prior to the arrival of the new millennium, executives at Ohara Memorial Hospital in Kyoto, Japan (see also Best Practices in Action, The Ohara Group, in Chapter 2), realized that one of the keys to the healthcare institution's success was progressive leadership that included M-oriented leadership behaviors described in Chapter 10 and an institutionwide quality initiative. This realization stemmed largely from the institution's goal to be an example of quality for both patients and employees.

This was the impetus for the introduction of quality circles throughout the organization. Following the introduction of the quality program's implementation, groups began to meet routinely to dissect operational issues and seek alternatives to processes that did not contribute to optimal service delivery. This was particularly useful in the foodservice department, where issues associated with the organization's rapid growth were prevalent.

As noted in the first best practice in Chapter 2, Ohara Memorial underwent dramatic changes that included requiring the foodservice department to deliver meals to patients in other buildings and to tertiary sites. Yet solutions regarding this and re-

lated issues were not immediately apparent. Thus, several quality teams were formed within the department, and each team was charged with a specific operational issue. The teams assumed responsibility to present management with solutions that produced quality outcomes and that were cost effective.

While this process took longer than expected, the results were very positive. Perhaps the best outcome, however, was one that was unexpected. The communication pathways that evolved from the within-group discussions, as well as from the heightened discourse between the teams and management, were both pervasive and long-lasting. In fact, they reportedly changed the way the department functions, as the lines of communication are now broader and clearer.

Another impressive benefit is the effect the communication within departments has had on the entire organization, resulting in a much better exchange of information among divergent departments. For example, if the foodservice department encounters issues that are compromising customer satisfaction related to nursing, a foodservice manager now speaks directly to the appropriate nursing supervisor. Such cross-departmental communication is widely accepted and is seen as a method of meeting or exceeding customers' expectations, since problems that would have otherwise gone unresolved can now be addressed.

This represents a dramatic shift from the traditional autocratic style that many expect from a large, well-established Japanese healthcare organization. Ohara, quite simply, is dedicated to finding new ways to solve old problems. The organization's leaders believe that anything that advances this goal is worth pursuing. This is the driving force behind the quality initiative and the new goal to continually improve communication efforts both inter- and intradepartmentally.

Today, quality is evident throughout the hospital. In addition, communication is one of the many ways that Ohara sets itself apart from other healthcare organizations. This leads to a higher level of care for patients and greater job satisfaction for employees. This is readily apparent in the foodservice department. As an observer noted about this department at Ohara, "Everyone is interested in how the patient feels. This doesn't just mean they want the patient to become healthy. They want the patient to be happy, too!"

Adopting a TQM approach to business is in no way akin to simply reworking production in the kitchen or changing service approaches in the front of the house. It is, when done correctly, a transformation in the way an organization manages its operations, which then transcends all areas. TQM involves focusing management's energies on the continuous improvement of all aspects, from food production to labor deployment, while at the same time leveraging the expertise of those who know how to improve things—the line employees. More than anything else, it results in changes in how work is done. Hence, TQM is both a mentality and a process.

Linking with the discussion from the previous chapter, TQM can also be applied to overarching functions, such as leadership in an organization. (In fact, this is an excellent place to start any TQM initiative.) To help illustrate how this might be accomplished, Figure 11.2 details a leadership framework.

The most obvious feature of this model is that it transfigures the role of leader, which usually occupies the top position in depictions of an organization's hierarchy, placing it instead at the center of concentric spheres of influence. The implicit message here is that the leader embraces informal authority, distributing leadership throughout the organization. And as evidenced by the larger area occupied by the outermost circle, the model underscores the importance of the stakeholders to the organization's success.

PICOS

Picos, derived from the Spanish word meaning "peak" or "pinnacle," was conceived by General Motors as a means to standardize and optimize the procurement process, involving vendors throughout the world. The method was originally aimed at helping General Motors' suppliers perform at their highest level, thereby reducing costs for both parties while improving quality. It was also viewed as a way to reduce waste and standardize different elements of manufacturing in the rapidly changing automotive industry.

The quality-improvement process evolved from the TQM paradigm, but with a fast-track application. The first 10 steps in the process are designed to prevent problems, and

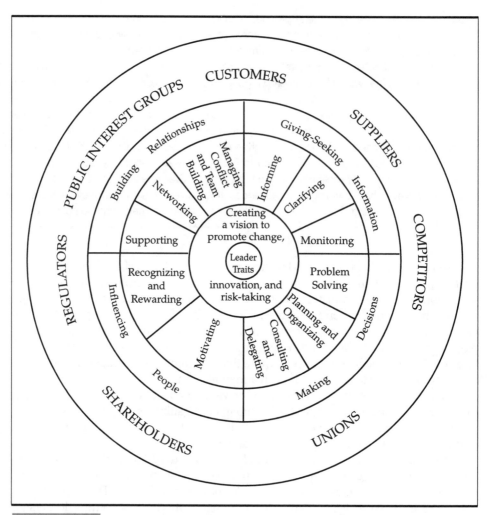

FIGURE 11.2 A TQM Leadership Framework. Reprinted from *Journal of Quality Management*, *1*(1), Puffer, S. M., & McCarthy, D. J., "A Leadership Framework for Total Quality Management," p. 113, copyright 1996 with permission from Elsevier Science.

the remaining 6 outline steps to correct problems. While organized in a logical sequence, some steps may play a greater role in certain applications, depending on the existing problems and the quality orientation. Moreover, if earlier quality initiatives have been used, some steps may be merged to reduce the complexity of the package.

For on-site foodservice operators, the basic Picos application begins with customer-satisfaction data, which are used to identify an area of improvement. A Picos team is then formed, consisting of supervisors, staff members, support people, and others representing a cross section of people whose jobs relate to the process in question. Both of these steps are typical of the TQM process described earlier. After that, however, the similarities begin to lessen. Following an intensive workshop, the team identifies both productive and wasteful elements of the process, brainstorms ideas for improvement, develops short-, medium-, and long-term action items, and then measures the ongoing progress. The fast-track approach is what is most impressive: Changes are made immediately following the identification of problem steps in the process.

The foodservice management team at Cayuga Medical Center has produced excellent results through the Picos approach to process improvement. Thus, these savvy operators are highlighted in this chapter's second best practice.

IMPROVING QUALITY IN FOOD PRODUCTION AND SERVICE: PICOS AT CAYUGA MEDICAL CENTER (ITHACA, NEW YORK)

Cayuga Medical Center at Ithaca, a 204-bed acute-care facility, has provided quality healthcare to the citizens of this upstate New York community for more than 100 years. It is one of the largest employers in the area, with 750 healthcare professionals along with an affiliated medical staff of 180 physicians. Cayuga Medical Center has seen many changes over the years and has sought to stay ahead of the competition by always focusing on quality in everything it delivers.

One of the tools the organization has adopted is the Picos approach to quality management. In the hospital's view, the three areas of peak performance are quality, customer service, and satisfaction. The organization uses the traditional Picos analysis methods—including waste, cycle time, and workflow—as primary avenues. Management is also strict about the vital Picos feature of aggressive implementation of solutions.

The organization's foodservice department was introduced to the Picos process when patient-satisfaction data indicated that there was a problem associated with food temperature; primarily, patients perceived that the hot food was not hot enough and, as a result, were dissatisfied with the dining experience. Foodservice managers immediately assembled a diverse team to attack the issue. The team consisted of individuals from the administration, community relations, nursing, and different areas of the foodservice operation including production, tray delivery, and clinical dietetic services.

Next, the team defined the focus and scope of the issue. The focus was simply patient satisfaction, with particular emphasis on serving the right meal at the right temperature with satisfying taste and presentation. The scope encompassed the interval from the instant the patient placed an order to the time when the meal was delivered.

The team then looked at the first area scrutinized through the Picos process: waste. Examining the entire meal-delivery chain from the customer's perspective, the team recorded 85 observations of waste. Examples included excess production, inventory that was not standardized (e.g., napkins, cups), and delays in the flow of information.

In examining cycle times, the team began by assessing ratios involving the time used for value-added steps as a percentage of the total cycle time and compared these with similar measures that integrated the time spent on non-value-added steps. In a related measure, they looked for workflow opportunities to reduce non-value-added steps. In doing so, they examined food-order placement, tray production, and the delivery of trays. While several tertiary opportunities were identified, the tray-production line—and problems stemming from it—was the most conspicuous.

By brainstorming the issues, the team generated 131 potential solutions. These ranged from using buffet-style service for the mental-health unit to changing the manner in which the foodservice department and the nursing staff communicated.

From the original list, the team—working together—identified 121 items that merited action. They then divided these items into general categories and further separated them into changes that could be initiated within 10 days, within 6 months, or within 12 months.

The last step was to set measurable targets. For example, they stipulated that the proportion of wasted trays—those that were delivered with the wrong food, too late, or to the wrong person—should drop from the average 1.6 percent to 1 percent in the short term to less than 0.78 percent in the long term (when all the changes had been instituted). Similarly, they identified specific targets for patient-satisfaction scores in the areas of food quality, temperature, and overall satisfaction.

It is important to note that many of the solutions were very simple. Yet, as is so often the case, they involved practices that had been in place for as long as anyone could remember. One good example is the configuration of the trayline, as depicted in Figure 11.3. Prior to use of the Picos process, the physical line was quite long, a remnant of times when more than a dozen employees worked the line at the same time. Also, the hot food was located at the beginning of the line. Through the quality-improvement process, non-foodservice employees suggested shortening the line and turning it around so that the hot food was the last to be placed on patients' trays. Simple? Of course. But no one would have thought of something so simple had they not sought the pinnacle of quality through a team approach.

FIGURE 11.3 Cayuga Medical Center's (Ithaca, New York) Redesigned Trayline, following One of the Picos Quality Initiatives. Copyright by Cayuga Medical Center at Ithaca Archives.

International Organization for Standardization

The **International Organization for Standardization** (ISO), a worldwide federation including some 140 countries, is a nongovernmental organization established in 1947. Its mission is to promote the development of standardization and quality on a global basis for organizations through the exchange of intellectual, scientific, technological, and economic resources. The organization is known as "ISO," not as an acronym but as a derivation of the Greek word *isos,* meaning "equal."

As part of its mission, ISO has developed sets of standards for quality management systems, which are accepted around the world. As such, these standards serve as both measures of quality and a framework for implementing an organizationwide focus on quality. The most commonly known standards are referred to as **ISO9000.** This family of standards includes ISO9000: Quality Management Systems—Fundamentals and Vocabulary, and ISO9001: Quality Management Systems—Assessment and Registration, among many, many others. Standards are constantly being added and revised, enhancing their thoroughness and utility. For example, the current 2000 system is markedly different from its 1994 predecessor in that it includes root-cause statistical analysis and proof of customer service and continual improvement, critical components that are part of a blended learning solution to maximizing quality.

ISO standards are important because an organization's compliance ensures that it has a sound quality-management system, which leads to reductions in customer complaints, reductions in operating costs, and increased demand for food and related services. Other benefits include better working conditions, increased market share, and increased profits. Another advantage is the growing trend toward universal acceptance of ISO9000 as an international standard, making it both a marketing and an operational tool.[3]

PROGRAM IMPLEMENTATION

As stated in the introduction to this chapter, quality initiatives can be highly successful—or they can share the fate of many a fad that management has perceived as a panacea for all known ills. In order to understand why some programs are highly successful while others die miserably (usually having a markedly depressive effect on morale as they flail about in the last stages of life), it is useful to look at the **key success factors**—those essentials that lead to an organization's triumph—and the common indicators that things are not working.

The first of these is complete buy-in by the entire management team. Since quality initiatives typically involve considerable change, even one member of the management team who does not embrace the idea can undermine it without even intending to. Line employees, like most people, are resistant to change; they think change means more work. Thus, it is management's role and duty to ease the process by providing enthusiasm and support. Without it, the quality quest will fail.

The second key success factor, which is linked to the first, is thorough and complete communication. Employees need to know what is happening before they can accept it. They need to know why a quality initiative is being introduced, what their roles are, and what the outcomes should look like.

This leads to a particularly vital key success factor, one that was discussed in a leadership context in Chapter 10: empowerment. The whole notion of most quality initiatives involves empowering employees to improve processes, particularly those processes in which they are specialists. A dish machine operator typically knows the efficiencies that are possible better than anyone else. If not empowered to integrate them into a workflow, however, he or she will likely not make the necessary changes.

The next factor involves clearly defined goals. What does management wish to attain? As mentioned in the earlier discussion of measuring quality, these goals should be substantive in the eyes of everyone involved. Simply stating that better quality service is desired will not produce the intended outcome.

The fifth factor is setting an appropriate pace for implementing the quality initiative. This is a function of the scope of the project, as well as of the number of areas that are under scrutiny at any given time. Changes that lead to better-quality food and service are not always simple and therefore take time. In addition, employees need time to consider the possibilities in shaping the outcomes.

The last key success factor is implementation for the right reason. In the past, many managers introduced quality programs because such managerial approaches were thought to be representative of progressive managers. This is not a sufficient reason. Moreover, introducing any quality program merely because it is popular too often fails to produce higher-quality output, alienating the employees who are most deeply affected by the related changes.

TOOLS FOR QUALITY MANAGEMENT

Owing to quality management's overarching focus, the quest for quality has resulted in mainstream applications stemming from a variety of disciplines that have robust applicability for on-site foodservice management. While entire volumes are devoted to the wide variety of tools available, the most relevant and those that offer the greatest utility in this industry include hoshin planning, the affinity process, and Pareto diagrams.

Hoshin Planning

Hoshin planning is a combination of strategic planning and policy deployment that integrates all levels of an organization. The term stems from the Japanese phrase *hoshin kanri,* which is translated as policy or target (*hoshin*) planning or management (*kanri*). Fundamentally, hoshin planning occurs at two principal levels: first, at the guiding or strategic planning level and, second, at the daily management level, pertaining to the more routine aspects of the operation.

Hoshin planning is an extension of the TQM process that allows an organization to plan and execute strategic organizational breakthroughs. The key is the identification and articulation of specific goals that cascade through the organization; these goals are based on the true capability of the organization. Like other TQM approaches, hoshin planning uses a continuous-improvement process such as the plan-do-check-act approach discussed earlier. It focuses on the key systems that need to be improved to achieve strategic objectives, and it requires participation and coordination of all areas in the deployment of resources. Finally, planning and execution must be based on facts and concrete objectives.

Here's how it works: Senior management, along with a heterogeneous pool of individuals including those at lower ranks, identifies the handful of key goals that are critical for the strategic advancement of the organization. From there, action plans are developed, each with a clear link to the aforementioned goals. Generally, the majority of the action plans involve lower levels of management. Hence, the next layer of management is brought in to develop action plans of its own that will serve to actualize senior management's plans. The process continues to flow to the lowest managerial level. Hoshin planning thereby creates clear action plans that share common elements related to the organization's strategic goals.

Affinity Process

The affinity diagram, or KJ method (after its author, Kawakita Jiro), was developed to aid in the discovery of meaningful groups of ideas within a raw list. It was not originally intended for quality-oriented management applications. Nonetheless, it has become one of the most widely used general management and quality-planning tools. In particular, the **affinity process** is often used as a starting point in hoshin planning but also serves as a useful tool in a variety of managerial applications.

An affinity diagram, which links common ideas, is most often used to refine a brainstorm into something that makes sense and can be dealt with more easily. An authority on quality approaches used in Japan, Kaoru Ishikawa, recommends using the affinity diagram when facts or thoughts are uncertain and need to be organized, when preexisting ideas or paradigms need to be overcome, when ideas need to be clarified, or when unity within a team needs to be created.[4] As one might expect, this is particularly useful for foodservice managers when different perspectives and frequent divisions of opinion are common.

In a nutshell, the affinity process involves everyone developing ideas independently. Through a series of clustering exercises, the separate ideas are brought together to form cohesive groupings. Thus, cognitive classification can be performed using disparate sources of ideas and creativity.

Pareto Diagram

More than a century ago, Alfredo Pareto (1848–1923) conducted extensive studies of the distribution of wealth in Europe. He found that there were a few people with considerable wealth and many people with little money. Today, this idea is a common element in economic theory. Perhaps of greater importance, however, others have adopted Pareto's analytic approach and applied it elsewhere.

With a focus on quality, Dr. Joseph Juran applied Pareto's technique to organizational processes and coined the phrases "vital few" and "useful many."[5] This is the origin of the 80-20 rule, which generalizes the Pareto principle to predict that 80 percent of the problems are the result of 20 percent of the component parts. In foodservice management this rule is often used to describe employee issues, where 80 percent of the problems are usually caused by 20 percent of the employees. As shown in Chapter 7, the rule can be applied more generally to explain inventory-cost issues.

A **Pareto diagram** is a graph that ranks data classifications in descending order, making identification of the vital few easier. In application to foodservice, these data classifications include causes of customer complaints, employee issues, reasons why overtime is used, and so on. The vertical scale can be dollars or frequency, depending on the data classification type. Note that a Pareto diagram is different from a histogram in that data classifications in Pareto diagrams are always categorical.

In Figure 11.4, which shows a Pareto diagram of customer-complaint classifications, it is clear that the vital few involve employee attitudes. Obviously, listing the categories in descending order could identify this extreme situation. The graph, however, provides visual impact, showing which vital few (or vital one in this case) deserve attention.

In practical terms, Pareto diagrams should dictate the focus of any quality-improvement program. It is sometimes tempting to solve the smaller problems. The greater return on investment, however, will come from using the analysis intelligently and seeking to remedy the problems of the vital few before broaching the challenges evident in other, less serious areas.

A second point is that Pareto diagrams are fluid: Once one problem is solved, the analysis must be performed again. Referring to the example highlighted in Figure 11.4, it is likely that the other issues will have greater representation once the problem of service is remedied. Granted, overall quality will increase dramatically by improving just that one area, but opportunities to continue the upward trend in quality improvement will remain available only if the process is continually revisited.

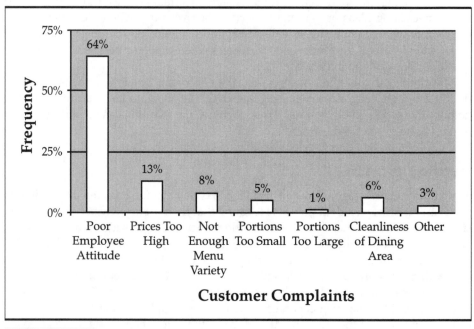

FIGURE 11.4 Pareto Diagram Using Types of Customer Complaints.

CHAPTER SUMMARY

Quality pertains to internal and external customers, basics such as food safety, and perceived value. In a service application, there is a subjective element to quality owing to the variance in perceptions among different users and constituents. Quality measurement falls into four categories: consumer-based measures, detection-based measures, financial-based measures, and process-based measures.

The most common and most useful of the several programmatic approaches to quality for foodservice managers are TQM, Picos, and the various ISO-related systems. Each approach has unique features, but all share several key success factors in order to produce the desired outcomes. These include buy-in from the entire management team, thorough communication, empowerment of line employees, well-defined goals, an appropriate implementation pace, and implementation for the right reasons.

Several tools exist that can help managers focus on quality. For foodservice professionals, the most relevant include hoshin planning, affinity processes, and Pareto diagrams. Hoshin planning is a blend of strategic planning and policy deployment, whereas affinity processes can help a management team refine a collection of ideas into something that can be used to shape organizational goals and objectives. Finally, Pareto diagrams provide a method for identifying key elements or processes that, because they have the greatest negative impact on quality, deserve management's immediate attention.

KEY TERMS

quality	continuous quality	ISO9000
customer-based measures	improvement	key success factors
detection-based methods	quality circles	hoshin planning
financial-based methods	Picos	affinity process
process-based methods	International Organization	Pareto diagram
Total Quality Management	for Standardization	

REVIEW AND DISCUSSION QUESTIONS

1. Using the definition offered for quality, what part is most subjective? Is this a good thing?
2. Which of the measuring mechanisms are most relevant to on-site foodservice management? Which one(s) are you most comfortable with and why?
3. Why has TQM received so much attention as a cure-all of operational ills?
4. Using a process common to on-site foodservice operations, describe the CQI process, including examples for each of the seven steps.
5. What makes Picos such a useful tool?
6. Why might a quality initiative fail? Describe the two elements that you believe are most likely to lead to failure.

7. Think of a goal that a senior management team might identify for a multiunit on-site foodservice operation as part of the hoshin planning process. How might this translate into an objective of the line managers? As part of your answer, describe the link between the overarching goal and how it might be achieved in part at the line level.

8. Think of an application other than the example given in the chapter for a Pareto diagram. Where would you find the necessary data? Using your example, what would you expect to find?

END NOTES

[1] A good reference for TQM, which addresses implementation in more detail, is Jablonski, J. R. (1991). *Implementing total quality management: An overview.* San Diego, CA: Pfeiffer & Company.

[2] For more on CQI, see Ryan, M. J., & Thompson, W. P. (1998). *CQI and the renovation of an American healthcare system.* Milwaukee, WI: ASQ Quality Press.

[3] As alluded to, ISO standards are numerous and complex. A good overview beyond that presented here can be found in Besterfield, D. H., Besterfield-Michna, C., Besterfield, G. H., and Besterfield-Sacre, M. (1995). *Total quality management.* Upper Saddle River, NJ: Prentice Hall.

[4] One of the better textbooks, which is also well translated, is Ishikawa, K. (1985). *What is quality control? The Japanese way.* Englewood Cliffs, NJ: Prentice Hall.

[5] While Deming is considered the father of TQM, Juran is seen as providing the fuel behind the global quality control movement. One of the best works by Juran is Juran, J. M., & Gryna, F. M. (1988). *Juran's quality control handbook* (4th ed.), New York: McGraw-Hill Book Company.

CHAPTER 12

SENIOR DINING: THE NEXT NEW FRONTIER

The mere mention of senior dining conjures up stereotypical images of "old people" eating bland meat loaf, mashed potatoes with the texture of rice paper, and over-cooked, anemic vegetables. The facilities, at least as propagated by yesterday's media, were fashioned after sterile environments that closely resembled psychiatric wards for the criminally insane. The dining areas were equally lifeless, with servers who had little passion for their jobs and little love for their customers.

Today, however, senior living and the associated senior dining segment of on-site service is a blossoming industry. Fueled by a growing percentage of Americans with gray hair, the segment is both dynamic and stratified, reflecting the different groups within the senior generation. While it seems contradictory to think of an industry that serves seniors as high energy in nature, this is in actuality exactly the case.

Accordingly, this chapter begins by exploring the aging population in order to provide a more thorough understanding of the segment's current and future markets. Next, the effects of these dramatic population characteristics are considered in terms of their potential impact on society at large. This leads to a discussion of the various senior living segments, followed by an exploration of how on-site foodservice providers can meet the needs in each of these arenas.

THE AGING POPULATION

At the beginning of the twentieth century, life expectancy in the United States was just under 50 years.[1] While people realized that this number would increase with increasing prosperity, even the most provocative futurists of the time did not predict the subsequent rate of increase. Even the U.S. government, when it enacted the Social Security Act in 1935 in part to provide "old-age assistance" and "old-age benefits," did so to protect the small group of individuals expected to live beyond 65 years of age. This is an important point, since life expectancy in 1935 was only 61.8 years. (See Figure 12.1 for a complete breakdown by decade and gender.)

At the onset of the twenty-first century, the average life expectancy of Americans was calculated to be around 77 years. This upward trend is expected to continue, with a similar rate of increase: In 2050, life expectancy in the U.S. is projected to be around 84 years. (See Figure 12.2 for a breakdown by gender.) Some suggest that there is a biological limit to this trend, arguing that no amount of good science will enable the human body to endure indefinitely. Nonetheless, for female children born at the

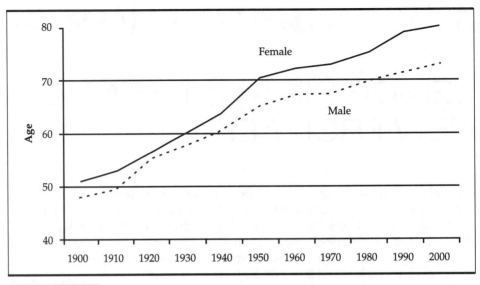

FIGURE 12.1 Life Expectancy in the United States by Gender, by Decade, 1900–1999. Adapted from National Vital Statistics System.

dawn of the current millennium, some prognosticators suggest that a great many will survive to see the twenty-second century.

These statistics lead to the more topical issue of how this changes the population composition of the United States in terms of age. For centuries, the number of older adults remained proportionately small, although the exact age ranges varied according to the relative life expectancies. In the past 70 years, however, mortality from almost all major diseases typically experienced at older ages has declined steadily. In addition, improvement in public health efforts, general medical technologies, hygiene, food abundance and quality, and general standards of living have been enhanced. This has led not only to the longer life expectancies described above but also, because of a concomitant decrease in births, to a growing proportion of the population that is older. For example, the population of people 85 and older is expected to increase by 33.2 percent between 2000 and 2010!

For younger "older" adults—those who have reached the once-standard retirement age of 65—the numbers are even more amazing. In 2000, the 65-and-older crowd represented 13 percent of the U.S. population. This number will slowly increase

Life Expectancy	2000		2050	
	Male	Female	Male	Female
	74.5	80.0	81.2	86.7

FIGURE 12.2 Life Expectancy in the United States by Gender, Current and Projected. Adapted from U.S. Census Bureau.

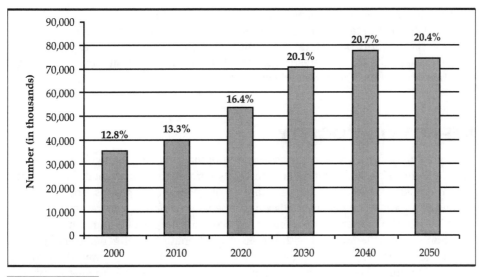

FIGURE 12.3 Number and Percentage of People over Age 65. Adapted from U.S. Census Bureau.

until around 2030, at which time it will exceed 20 percent. (See Figure 12.3 for a breakdown by decade.) These percentages are even more amazing when we consider that in 1860, half of the nation's population was still under age 20; in 1950, half of the population was under age 30. According to reasonably conservative projections, by 2030 half of the population will be 39 or older.

Is the United States unique in terms of having a growing older population? In the more developed countries, it would appear that higher life expectancies are creating the same situation. As shown in Figure 12.4, which includes the five countries with

	Life Expectancy at Birth	Percentage of Population Aged 65+
Worldwide	67	7
More developed	75	14
Less developed	64	5
United States	77	13
United Kingdom	77	16
Germany	78	16
France	79	16
Japan	81	17

FIGURE 12.4 Life Expectancies and Percentage of National Population Aged 65+ by Country, for More and Less Developed Countries and Worldwide. Adapted from 2002 World Population Data Sheet, *United Nations Demographic Yearbook,* 2002, and *Population and Vital Statistics Report,* UN Statistics Division.

today's highest average life expectancy, the United States is nowhere near the top. Using 2000–2002 data, Japan currently has the largest number of seniors as a percentage (17 percent) of its total population as well the highest life expectancy (81 years). Perhaps the most interesting aspect of these data is the difference among the countries shown and the worldwide averages.

IMPACT ON SOCIETY

This "graying of America" and the associated population shifts, particularly those evident in Figure 12.3, will result in an interesting combination of societal changes. First and foremost are the changing delineations of generational groups. The associated **generation gaps**—the perceived disconnect between different generations—have always been part of society. The second is the economic impact. Owing to unequal distribution of wealth combined with the disproportionate size of generational groups, the spending habits of older Americans will play an increasingly important role in business ventures. The final societal change that results from the growing number of older folks pertains to family settings and living arrangements.

Generations

Misunderstandings between generations are nothing new. This generation gap, however, has traditionally applied to the differences between parents and their adolescent children. While this will probably never change, the growing number of older citizens makes differences among multiple generations more apparent.

In order to understand this more fully, it is useful to describe each of the generations that exist today.[2] The first, and oldest, group is referred to as the **GI generation.** Born between 1900 and 1924, this generation grew up with the Depression and went on to fight in the two world wars.[3] The **silent generation,** born between 1925 and 1945, came of age too late to be war heroes but were taught the lessons of the Depression by their parents and understand the tragedies of war from their own youthful observations. **Baby boomers,** born from 1946 to 1964, were the result of the prosperity and hope for the future that sprang from the conclusion of World War II; this is the largest of any generation to date. **Generation X** refers to those born between 1965 and 1980; the "X" refers to the unknown characteristics of the generation at the time.[4]

The differences among these generations, even using the most basic anecdotal evidence, are substantial. Managers already understand the complexity of managing a workforce that at times is composed of members of all four. These differences, however, will become more prevalent from a societal perspective as the baby boomers approach retirement age.

Economics

How will each of these generational groups affect the economy in the near and distant future? For example, the number of remaining members of the GI generation with incomes sufficient to afford some sort of living assistance is striking; in 1999, according to data from the Assisted Living Federation of America, over 57 percent of people over age 80 had annual incomes topping $15,000 and over 38 percent had annual incomes of at least $25,000.

Similarly, the silent generation, while relatively small in number, today possesses an inordinate amount of wealth. Thus, this group has a substantial influence on the economic enterprises offering products of interest to its constituents. This explains the emergence of upscale retirement communities that cater to this group.

Carrying this example into the future, it is easy to understand the effect that the baby boomers will have. Today, members of the baby boomer generation dictate many consumer trends owing to the group's substantial disposable income. These buying patterns have influenced every major industry, including hospitality. With the oldest boomers quickly approaching age 65, these buying decisions will provide incentives for business enterprises to cater more fully to a growing older segment of the marketplace. And as the bulk of the generation moves into the retirement phase, their economic influence in concert with their sheer numbers will continue to shape many of the related market economies.

Societal Changes

The aging population has other effects beyond those dictated by economic factors. One is a change in attitude regarding retirement age. While the median age at retirement fell from 74 in 1910 to 63 in 1999, there is a growing reverse trend for Americans to continue working for pay after turning 65. The phenomenon, in fact, is more prevalent in the United States than in other wealthy nations. While the trend will likely be influenced by changes in Social Security that extend the official age of retirement to 67, it is also fueled by the need for people to work longer in order to save enough to carry them through their extended lives.

In terms of senior living, another major change is the decreasing proportion of widowed women seeking suitable accommodations. As both genders' life expectancy increases, more couples will look for living situations that match their combined preferences for housing and foodservice. On the other hand, a growing percentage of the older population will be divorced. As of 2000, only 7 percent of people over age 65 were divorced and had not remarried, whereas 15 percent of women and 12 percent of men aged 55 to 64 were in this same category.

Consider also the number of people who need assistance with **daily living activities,** such as bathing, dressing, and eating. According to 2000 census figures, about 6.5 million older Americans need assistance with daily living activities. By 2020, this number is expected to double.

Finally, and perhaps of greatest importance, the increasingly larger proportion of older people will likely experience greater disparity in terms of the distribution of wealth. Specifically, it is probable that the distinction between those with money and those without it will become much more pronounced. This disparity will also put unprecedented demands on the remaining population to care for its elders. In addition, products such as housing and dining options will need to include multiple segments and price structures in order to meet varying needs and economic resources while also providing a high level of care to everyone, regardless of financial status.

SENIOR LIVING SEGMENTS

Not surprisingly, the aging population and the aforementioned changes in society have created an increased need for comfortable housing options along with varying levels of healthcare and foodservice needs. But before embarking on a discussion pertaining to matching foodservice options with these customers' preferences, it is necessary to outline differences in the various senior living segments. While going by a variety of names, depending on the region, these five general segments include independent living, assisted living, skilled nursing, hospice and home care, and the multisegment continuing care retirement communities.

Independent Living

Independent living is a housing option that most closely resembles private residences such as those found in most suburban neighborhoods. In essence, independent living centers provide a retirement setting where the community is based on members of similar age who are in good health. As in any communal setting, the activities focus on common interests.

Foodservice varies, but in most independent living centers at least one meal per day is available. Meal plans also vary and sometimes are quite similar to those offered in colleges, where residents can select plans based on the number of meals they want each month. In many independent living centers, the main dining rooms are extremely elegant and may be open to the public as well.

Independent living centers do not provide any healthcare options and are not equipped to provide assistance with daily life activities. (See Figure 12.5 for a graphic rep-

	Daily Life Activities (e.g., dressing, bathing)	Resident's Overall Health (physical and emotional)	Community Activities (social events, outings, etc.)	Environmental Services (laundry, housekeeping)	Health Services (nursing care, medication, etc.)	Environment (in terms of personal freedom)
	Assistance none → complete	*General health* poor → good	*Number & type* none → many	*Number & type* none → many	*Level of care* low → high	*Restrictions* many → none
Independent Living	●○○○○	○○○○●	○○●○○	○○○●○	●○○○○	○○○○●
Assisted Living	○○○○●	○○●○○	○○●○○	○○○○●	○○●○○	○○●○○
Skilled Nursing	○○○○●	●○○○○	○●○○○	○○○○●	○○○○●	●○○○○
Hospice and Home Care	○○○○●	●○○○○	○●○○○	○○○○●	○○○○●	●○○○○
Continuing Care Retirement Communities	●●●●●	○●●●●	○○○●○	○○○●●	●●●●●	○○○○●

FIGURE 12.5 Characteristics by Senior Living Segment.

resentation of the services and situational factors for each type of housing option.) While not considered healthcare centers, many independent living settings do serve as the first step in the matriculation to housing that provides some types of medical services.

Assisted Living

Picturing the options for senior housing on a continuum, the next level of care is an **assisted living** center. Assisted living is a combination of housing, personalized support services, and healthcare designed to meet the needs of those who require help with daily living activities. Among the services typically provided are housekeeping, laundry, transportation, and, of course, assistance with eating, bathing, toileting, and so on. Many assisted living centers also offer specialized services for those suffering from Alzheimer's disease or varying forms of dementia.

One primary feature of any assisted living center, regardless of its size or resident composition, is its foodservice, which provides three meals per day for all residents. These may be delivered via tray service or served restaurant-style in a central dining room. Some assisted living centers with a majority of residents who do not require assistance also offer buffet-style service to allow for more independence in menu-item and quantity selection.

Skilled Nursing

An elderly person who requires 24-hour nursing care often chooses a **skilled nursing** facility. As the name indicates, skilled nursing centers provide nursing care for patients requiring both medical treatment and assistance with activities of daily living. Residents are constantly supervised by a team of licensed nurses, therapists, social workers, dietitians, and certified nursing assistants.

In cases where a resident is convalescing, a skilled nursing facility may function as a rehabilitation healthcare center, helping the individual heal. In other cases, the goal is to provide healthcare, as well as fill recreational, psychosocial, and spiritual needs indefinitely. The key is to transcend the role of healthcare provider, providing a place where people can enjoy a high quality of life for extended periods.

Foodservice in skilled nursing facilities is similar to that found in assisted living centers but is tailored to provide a much greater number of specialized meals and snacks. This type of foodservice requires substantial expertise in terms of both food production and dietetics. In addition, a sophisticated level of foodservice management is required to ensure that residents receive both nourishment and satisfaction.

Hospice and Home Care

Hospice care and **home care** are additional options. Home care is provided in an individual's home by a professional healthcare provider and aims to keep the individual functioning at the highest possible level. Hospice care is a combination of facility-based and home care provided to benefit terminally ill patients and support their families. The goal of both types of care is to provide high-quality medical care and reduce pain and suffering while preserving the highest possible quality of life.

Foodservice in this area varies considerably and depends largely on the level of service, type of service provider, and needs of the person receiving care. For some, eating is not pleasurable due to medications and associated interactions. For others, meals remain an important part of the day.

Continuing Care Retirement Communities

A **continuing care retirement community** (CCRC) is a true community that offers several levels of assistance, including independent living, assisted living, and skilled nursing care. Again referring to Figure 12.5, CCRCs offer a broad spectrum of services and are intended to provide housing for individuals as their healthcare requirements change. CCRCs emphasize the importance of aging with dignity by proving supportive environments that are easily tailored to a broad range of people.

Many CCRCs emulate the atmosphere of a country club. Features may include multiple golf courses, spa facilities, and numerous restaurants. Nested within this elaborate setting, however, is also a healthcare center that provides care for those who need it. For example, it is not unusual for CCRC residents to enjoy the independent living setting for a number of years before moving into the assisted living part of the community when their physical condition requires it. It is this continuity that appeals to many.

Foodservice in CCRCs, as expected, is a multifaceted endeavor. It is common for the level of service and menu offerings in the dining room to rival those of upscale restaurants. Yet the same kitchen that supplies the filet mignon and shrimp scampi also prepares meals for the assisted living center. Therefore, the foodservice managers must deliver on a large number of implicit promises; residents who move from the independent living area to the assisted living center may realize that they can no longer eat some of the rich entrées they formerly enjoyed, but they nonetheless will expect the same level of quality and service.

HOW FOODSERVICE CAN MEET THE SENIOR MARKET'S NEEDS

Providing foodservice that meets this seasoned market's expectations is extremely challenging. First, this group knows what it likes. If a person has expressed a lack of interest in eating any kind of seafood for the past several decades, it is unlikely that this preference will change.

Another challenge relates to the physical reality that the senses of taste and smell diminish over time. In fact, some research indicates that some 75 percent of people over the age of 80 experience difficulty perceiving and identifying various tastes and smells. On a related note, most adults have 206 taste buds; studies have shown this number to drop by more than half after age 75. Thus, it is not uncommon for an older individual to find food bland and tasteless. This puts the onus on the foodservice team to create meals that live up to memories while also ensuring that the meal is not overly laden with salt or fat, which might compromise the customer's health over time. (This is not to say that a person who has eaten fatty foods should be expected to switch to a low-fat diet just because he or she has moved into a senior living facility. It is the foodservice provider's responsibility, however, to offer menu items that are both delicious and nutritious.)

Still another issue pertains to the feeling of independence that is sometimes compromised when people leave their homes in exchange for a senior living option. The problem begins with the decision itself; too often it is the children or sibling who makes it, not the person who moves into the new setting. In order to foster as much independence as possible, it is critical that all residents have as much control as possible over what and when they eat. While this is difficult when the resident needs help eating or making menu decisions, it is still a vital responsibility that leads directly to quality-of-life issues.

So, despite these numerous issues, what can the on-site foodservice manager do for this segment? Among the opportunities for rising above others in this area are giving people what they want, offering novel options, and using automation to maximize efficiency and enhance the dining experience.

Let Them Eat Cake

As mentioned throughout Part II, the only way to exceed customers' expectations is to know those expectations. Most foodservice directors in senior living centers have weekly or monthly meetings with a committee of residents that attempts to speak for the majority. Too often, however, the manager receives information that is representative of only the most outspoken individuals. The solution is to take the extra step by collecting information from differing subgroups within the population. This is not to say that it is possible to make everyone happy all the time—that simply will not happen. But it is possible—with ample information in hand—to make decisions about menus, as well as service style, that will produce overall positive outcomes.

The trick here is to use creative ways to get information. It is not enough—it's not even close—to simply walk up to a table in a CCRC dining room and ask for feedback about the meal. Too often customers will alter their response on the basis of who is joining them, or they may not give any useful information. This does not mean that the dining experience is all that it could be. Some managers have used alluring technology to gain residents' input, such as a computer with a survey in the main lobby. Other approaches involve games in which residents must reveal their favorite foods or dishes in order to compete.

Reliable information about what residents like will become even more important in the coming years. It is likely that most senior housing facilities will have more than one generational group represented in the senior living center. And within these generational groups will be multiple ethnicities as well as differing regional representation from throughout the world. Without sufficient data, the foodservice team will be shooting in the dark while working to achieve the impossible.

More Than One Way to Skin a Cat

More often than not, a foodservice manager does things as they have always been done. This is not always bad, particularly as it relates to integrating control systems or motivational techniques. In terms of delivering unprecedented levels of quality food and service, however, it is often vital to stir up the pot.

At most upscale CCRCs, for example, foodservice professionals focus on making the dining experience as upscale and polished as possible. After all, this is what people expect and what—in the case of prepaid meal plans—they are paying for. There are times, however, when it is appropriate to offer something different. This might be a dinner with a casual theme where a coat and tie are not allowed and the servers wear shorts. It might be a luau where the ceremonial pig is authentically cooked in a hand-dug pit. Or maybe it's an add-on feature, such as an ice-cream sundae bar following a formal dinner.

Some managers are using the mix of population in their facilities as inspiration for different types of specials, leveraging the tried-and-true practice of offering different ethnic cuisines. The possibilities are endless, and the opportunities are as diverse as the resident population. Another option, which may appeal to some groups more than others, is dining during nontraditional times. Some residents might prefer a brunch-style meal rather than lunch and dinner. If the population is large enough to make staffing such an additional meal cost effective, this can provide a very satisfying option for some residents.

Other options for delivering foodservice in unique ways, many of which require substantial capital improvements, also have considerable utility. For example, adding a casual dining outlet, modeled perhaps after a 1950s-style malt shop, can provide a wonderful dining option. And because the food and labor costs for such a concept are substantially lower than those for the main dining room, it is likely that by using the appropriate amortization schedule, the outlet will pay for itself. Another idea is providing additional private dining rooms for use by families and small groups. Again, this affords the foodservice operator the opportunity to offer a wider range of services.

Maximizing Efficiency

As most experienced managers know, there are always better ways to do things when the proper systems or tools are in place. In skilled nursing centers, for example, the monitoring of resident intakes is always a challenge. Similarly, finding ways to ensure that the right resident gets the right meal can be daunting, particularly when the size and scope of the facility preclude constant supervision of meal delivery.

Several options can help. The key to any approach is that it should maximize value to the resident or increase efficiency (and correlatively reduce costs); in the best cases, it does both. An operator who found just such a system is this chapter's best practice.

BEST PRACTICES IN ACTION

OPTIMIZING AND SIMPLIFYING OPERATIONS IN SENIOR DINING THROUGH TECHNOLOGY AT THE PETER BECKER COMMUNITY (HARLEYSVILLE, PENNSYLVANIA)

The challenges discussed throughout this chapter paint an accurate picture of the difficulties foodservice operators face in serving the senior dining segment. Residents' needs are numerous and very specific; they also have high expectations regarding foodservice. Finally, the need to minimize the "institutional" quality of the food and related services is critical to delivering the utmost in dining experiences.

The Peter Becker Community, a CCRC with some 350 residents (and plans to add housing for 500 more in the near future), has successfully embraced cutting-edge technology to deliver an ideal combination of food-related services to a diverse resident population. This population includes 94 residents in a skilled nursing center, 34 in a personal-care center (akin to assisted living, discussed earlier), and the remainder in apartments featuring full kitchens and deluxe accoutrements. Those in the independent residences are very active and take pride in both preparing their own meals and dining in the upscale facility within the community.

After years of doing everything by hand, such as creating tray tickets—the cards that list each resident's preferences and dietary restrictions—and manual inventory sheets, the foodservice management team realized that the problems and inefficiencies associated with a manual system were too numerous to combat without embracing technology. To that end, they integrated GeriMenu, a foodservice software package designed specifically for healthcare by CBORD. While many senior living centers use various components of GeriMenu, the Peter Becker Community has harnessed all components of the package, including the advanced modules that integrate specific hardware.

From a management perspective, the technology has facilitated the automated production of tray tickets complete with pictures of residents (as shown in Figure 12.6). Foodservice employees no longer have to worry about illegible information, the result of smudged ink, or inaccurate tray deliveries because a new employee didn't know residents by name. Snack labels are produced by the system, making delivery of snacks a simple task. In the main dining room, the system prints restaurant-style menus, contributing to the professional delivery of foodservice.

In the back of the house, the system automatically issues production sheets corresponding to each day's menu. Costing information by resident, including meals, snacks (commonly referred to as "nourishments"), and supplements, is generated and is accurate to the penny. Among other attributes, the system even generates birthday lists so that the foodservice team can plan ahead and deliver special celebratory food items for every single resident on this momentous occasion.

As implemented, the GeriMenu system at the Peter Becker Community also supports their clinical

E-106	**Stroud**	Mrs. Gertrude	Today's Date: **9/3/XX**	**SUPPER**
PLATE GUARD REQUIRED			**Regular Diet**	

	Sustacal, Van.	5 Oz.
HALF	Sauce-Tartar	1 Pkg.
	Butter	1 Tsp
	Sugar 2/s & P	2 Sug,s,p
	Dressing-Italian	1 Pkg.
	Triple Bean Salad	1/2 Cup
	Peach Halves-njp	1/2 Cup
	Roll-Dinner	1 Ind.
	Lettuce Salad	1/2 Cup
HALF	Fish-Fried	4 Oz.
	Broccoli-Seasoned	1/2 Cup

PhotoCheck™ by GeriMenu

FIGURE 12.6   An Automated Tray Ticket Produced by the GeriMenu System.

efforts in varied ways. According to the dietitians, the automated calculations, including those pertaining to food intake, make life much easier and allow them to spend more time with residents. The graphic presentation of weight changes by resident is another tool that helps them monitor residents' health. Finally, the system tracks review dates pertaining to diet orders, chart reviews, and nutritional assessments, all critical to maximizing resident care.

The management team also has saved considerable time by using the inventory management tools. The system facilitates manual inventory recording through the use of hand-held computers that feature lists of food items corresponding to their location in the kitchen. The data, which automatically populate existing databases where current pricing information resides, are then integrated automatically to produce weekly cost reports, inventory valuation information, and various other key informational reports. Moreover, the efficiencies afforded by the system, in addition to helping managers manage better, have also resulted in reduced labor expenditures. For example, the time needed for the physical inventory has been reduced by some 70 percent.

As the foodservice director noted, most of the tasks that made the job difficult and most often resulted in dissatisfaction for residents are now performed through the integrated technology. Thus, lower labor costs, better service, more time available to concentrate on residents, and better control of the financial aspect of the operation are optimal now. The idea of doing things the old-fashioned way just seems, well, old-fashioned.

The possibilities for feeding this growing segment of the population are enticing. They also come with a unique set of challenges. Some of these will be overcome with technology, others with novel production methods in concert with efficient means of service delivery. Regardless of how these opportunities are addressed, one thing is clear: It is incumbent on all foodservice providers to ensure that this sizable customer base is satisfied.

This market is composed of those who most recently blazed the trail for all of us. As service providers, then, we have a unique responsibility. After all, if we, as foodservice operators, cannot contribute to helping people age with grace and honor by providing them with the opportunity to enjoy the pleasures of fine food and service, then who can?

CHAPTER SUMMARY

Life expectancy today and in the coming decades continues to follow an apparently limitless upward trajectory. In addition, because of fewer births and the aging of a large proportion of the population, the percentage of older people in the United States is expected to increase, spiking substantially in 2030, when people over age 65 will represent more than 20 percent of the population. This trend is also seen in other wealthy countries.

In order to understand the potential impact of these changes on society, it is necessary first to appreciate some of the differences among the generations. In large part, these differences stem from the different factors that have influenced each generation, such as wartime and national prosperity. These differences among generations mean that each one creates unique economic influences as it ages; this is particularly true for the boomer generation, which is the largest. As each group ages, we will also witness changes in family settings and living arrangements.

The increased number of older Americans has also created the need for a more stratified senior living industry wherein each segment is targeted at a specific group. These different types of senior living arrangements include independent living, assisted living, skilled nursing, hospice and home care, and the multisegment CCRCs. Not surprisingly, each has its own needs and nuances in terms of foodservice.

In order to meet this changing market's needs, both now and in the future, foodservice managers in the senior living segment must ensure that they are delivering what their target market wants; this is true of any segment but is particularly problematic given the heterogeneity of this particular market. It is also critical to offer novel options and use automation and technology to maximize efficiency and enhance the overall dining experience for everyone involved.

KEY TERMS

generation gaps
GI generation
silent generation
baby boomers
generation X

daily living activities
independent living
assisted living
skilled nursing
hospice care

home care
continuing care
   retirement communities

REVIEW AND DISCUSSION QUESTIONS

1. Why is the number of older Americans growing?

2. As stated in the chapter, some experts predict that average life expectancy cannot continue to increase because of the body's physical properties. What would a scientist in 1900 have said to this? What will a scientist in 2100 say?

3. Generation gaps are apparent in a variety of settings. Using your own experience, describe where a generation gap was apparent in a workplace setting.

4. If a growing number of seniors cannot afford senior living in the future, how will that affect other generations? Has anything like this happened before in this country?

5. What is the biggest challenge in providing people with assistance with daily living activities? Will this change in the future?

6. Independent living centers are not intended to provide healthcare. When occupants become ill, therefore, they must leave. Many times, their rooms or cottages are immediately rented out to others. Thus, when they become well, they must wait for another to become available. What are the psychological factors that might affect residents as a result?

7. CCRCs offer many advantages, not the least of which is the ability of residents to remain even when their health deteriorates. Why doesn't everyone choose this option?

8. Outline three options that a foodservice manager might employ to increase customer satisfaction in an independent living setting. Would any of these be viable in an assisted living center?

END NOTES

[1] Unless otherwise noted, statistics cited in this chapter were generated by the U.S. Census Bureau.

[2] Seminal work in this area can be found in Strauss, W., & Howe, N. (1991). *Generations: The history of America's future, 1584 to 2069*. New York: Morrow and Co.

[3] The precise determination of cutoffs for each generation varies among researchers, but those listed here are fairly common.

[4] Regarding generational differences, it is important to note that no one generation is better than another. Regardless of birth date, each generation faces challenges of poverty, of prejudice, and of cultural displacement. Nonetheless, there is a tendency—one that defies logic and evidence—to position one generation as being better than another. For example, it is easy to treat with greater reverence a generation that is credited with more heroic accomplishments than one that is remembered for self-indulgence and moral compromise. Regardless of the truisms, such prejudicial generalities are counterproductive.

CHAPTER 13

MULTIPLE SERVICES: IS FOODSERVICE ENOUGH?

As the world of on-site foodservice has grown in complexity, many managers have looked beyond the walls of the foodservice department in an attempt to gain greater synergies and, with them, efficiencies, economies of scale, and enhanced service delivery. At first, such ideas were met with ridicule. In a B&I setting, for example, people questioned the logic of hiring an employee to clean offices before dawn and then serve breakfast in the cafeteria. In healthcare, naysayers thought it disgusting to envision an employee delivering a meal and then cleaning the same patient's bathroom.

Managers with vision, however, realize that there is potential in the idea of managing multiple services as a suite of hospitality applications. Some easily visualize the potential in training people not just about departmental duties but also about service, and only then differentiating how service might be provided (e.g., foodservice and housekeeping). Others visualize the possibilities of sharing labor within related departments and how that might contribute to a better work environment, particularly during times of limited labor availability.

This chapter, then, addresses the subject of multiple services. It begins with a conceptualization of the potential synergies resulting from managing more than foodservice alone. Next, the chapter discusses different management structures, one of which is artfully demonstrated in the associated best practice. Finally, we focus on the cross-utilization of labor that is made possible by managing multiple services. Again, a best practice illustrates that this is not only possible but also highly beneficial.

SYNERGY IN SUPPORT SERVICES

At one time, the lines separating support services such as housekeeping and foodservice were clearly delineated. Whether in schools, B&I, healthcare, or any other on-site segment, managers and employees were hired and trained within the confines of a specific service. It didn't much matter that the foodservice employee scrubbed the dining-room floor exactly the way the housekeeper cleaned the lobby floor. These were treated as separate processes managed by separate people. And, more often than not, the employees in one group perceived themselves as enjoying a higher job status than those in the other group.

Sometime in the mid-1980s, however, people started to rethink this arrangement. Driven largely by managed-services companies that preached the advantages of grouping multiple services under one umbrella (which they would manage), word began to circulate among the various on-site segments. Clearly, there are advantages to this arrangement. There are also some inherent challenges.

Advantages

The advantages stem from the unconstrained efficiencies and cost savings possible when an organization moves from a silo approach to a more organic, seamless organizational structure that includes combined support services. The first—and perhaps most obvious—advantage is the savings in labor dollars achieved by combining management positions. Instead of a highly paid group of department heads, an organization can appoint a single general manager and utilize lower-paid supervisory staff within the different areas.

The next advantage is the single-mindedness that leads to better service delivery. Consider a catered event within a host organization. In the traditional management structure, the food would arrive, but the room might not be clean because the housekeeping staff wasn't aware of the event. While this could happen even when services operate more symbiotically, the likelihood of such an oversight is greatly diminished. A related plus is shared knowledge regarding all cross-departmental activities. Taken to the next level, tasks that affect multiple areas can be accomplished more easily under a single manager.

The final and most potentially substantive advantage—the one with the most dramatic impact—is the sharing of employees among the different support services. Does this mean that a manager can send the dish-machine operator to fix an air-conditioning unit? Of course not. But it does mean that shortages of line employees can be offset by backfilling from other areas. This, in turn, leads to better understanding of what goes on in the organization on the part of employees. It also creates a team environment that, as discussed in Chapters 10 and 11, leads to a more productive workplace. Finally, the crossover of functions creates the opportunity for job enrichment, which in turn leads to higher job satisfaction.

Challenges

Is combining services a bed of roses, with no downside? In truth, it is a difficult process, one that requires planning and continuous review. It also can create uncertainty for those involved, particularly those line employees who have known only the traditional management approach during their tenure.

According to those filling multiservice management positions (also known as "general managers"), the biggest challenge is maintaining the delicate balance of managerial functions that must take place. For example, most general managers find that the number of meetings is sometimes overwhelming. This, in turn, creates the need for unprecedented **time management.** Unfortunately, time-management skills of this magnitude must be learned and are not a part of every manager's repertoire.[1]

The other challenge for general managers is separation from the line employees. When managing only foodservice, for example, a manager may have daily opportunities to work with line employees. These exchanges lead to better communication and trust, eventually creating a workplace environment that is conducive to maximum productivity. Upon assuming greater responsibility, however, general managers may find that they have few opportunities to work shoulder to shoulder with employees, regardless of the department. Granted, the manager may be able to effect greater global changes that eventually lead to a better environment for the employees. From the employees' perspective, however, the reality is that the manager is not often around, which often creates resentment and ill will.

MANAGEMENT STRUCTURES

The notion of combining two or more support services, as alluded to earlier, requires considerable forethought given the different management structures that are possible. To what extent does the organization want to depart from the traditional organization chart or reconsider how departments are combined? And who should manage this area if multiple services are combined? A manager's success may depend on attributes uniquely suited to the multiservice environment. Without the proper talent, this innovative approach is doomed from the start.

Departures from Tradition—Multiple Possibilities

In Chapter 2, we discussed the traditional reporting relationship within a host organization for the foodservice director, using an example from a hospital (see Figure 2.3 on page 34). This structure was not created with the customer in mind or with the goal of maximizing efficiencies. In most cases, it evolved over time and in some instances may no longer even fit the needs of the organization. The problem is that the changes required for improvement can be so monumental that most companies elect to make the most of things as they exist. This is particularly true for the configuration of support services, since the person to whom the multiple departments report rarely has a deep understanding of all of the related disciplines.

This is where combining multiple hospitality-related services is most exciting and challenging. Since they are unrelated to the organization's core business, rearranging these departments has little direct impact on the greater organizational concerns. This, then, allows for experimentation and vast departures from tradition.

In most cases, the most logical approach is to combine foodservice and housekeeping. As noted earlier, many of the tasks already overlap. In addition, the entry-level positions used in both areas create opportunities for shared labor. Both areas also provide opportunities for line employees to interact directly with customers, thereby creating shared experiences that can be leveraged accordingly.

Another logical addition is laundry (for those organizations that require such a service). In a factory, for example, the laundry function may already be a part of housekeeping, making the combination even more logical. This combination also mirrors the way many hotels group services, wherein laundry and housekeeping share employees, office space, and so on.

Facilities management, which operates under various titles such as "maintenance," is yet another support service that it makes good sense to group with other hospitality-related services. Too often a receptionist needs to call housekeeping to clean a spill, foodservice for coffee for the boardroom, and maintenance to replace a light bulb; this is typical in an organization where each department has sole ownership of specific types of tasks. There are few reasons, however, why such functions could not be performed by one entity.

Other areas that may or may not be appropriately combined with the core hospitality services include communications, security, conference services (such as room scheduling), fitness-center management, parking, grounds, reprographics, and—for operators in healthcare—patient transportation. Again, each organization is different, and each has a specific set of support services that relate to the product or service that the host organization provides. The key is to combine those services that share as many functional commonalities as possible and that lead to opportunities to share resources, particularly employees.

An example of how one organization grouped its key support services and realized efficiencies as well as savings in labor dollars is described in this chapter's first best practice.

A UNIQUE MANAGEMENT STRUCTURE AND NOVEL APPROACH TO TRAINING PRODUCES RESULTS AT BETHANY HEALTHCARE CENTER (FRAMINGHAM, MASSACHUSETTS)

Bethany Healthcare Center, a 169-bed assisted living center in Framingham, Massachusetts (see Figure 13.1), is unique in a number of ways. For example, a majority of the residents are nuns. The building itself is quite old, although the interior is attractive and in good condition. The rooms, while well adorned, are small by modern standards.

There are two more intriguing features, however. The first is the way support services for the healthcare facility are organized. At Bethany, the foodservice, housekeeping, laundry, and maintenance departments all fall under a common umbrella and are managed by a single individual. Does this mean that Bethany is management heavy since each area requires dedicated management? No. In fact, when the center consolidated the services into a hospitality department several years ago, they actually reduced the number of managerial and supervisory FTEs.

Under the latest iteration of the management structure (see Figure 13.2), one operations manager is responsible for foodservice, while another operations manager oversees both housekeeping and laundry. Both individuals report to the general manager, who also has day-to-day responsibility for managing maintenance. The general manager maintains the financial records and also conducts walk-throughs of all the areas on a weekly basis. He also ensures that communication among the areas is free-flowing and frequent.

This arrangement is somewhat unusual in that there is still some autonomy within each area. It dispenses with the organizational boundaries common in other organizations, however, owing to the reporting relationships. Moreover, there is still a sense of belonging to the "hospitality services" department that extends beyond the identity employees form within their individual areas.

One advantage, according to those who spend the majority of their waking hours under Bethany's roof, is that communication among the four areas has increased, resulting in minimized re-

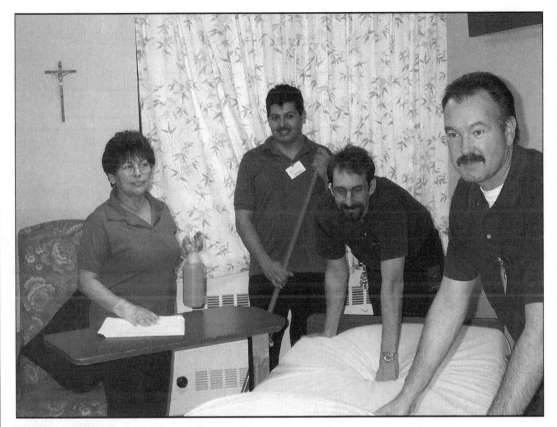

FIGURE 13.1 Hospitality Workers at Bethany Healthcare Center, Two Performing Maintenance Functions and Two Performing Housekeeping Duties. Copyright by Bethany Healthcare Center.

work. In the past, housekeeping sometimes stripped and waxed an area, only to have maintenance ruin the finish the next day while relocating a piece of equipment.

Another advantage is that the organizational structure results in more uniform delivery of services. Since each area is knowledgeable about what is happening elsewhere, labor shortages no longer have a dramatic effect. If the number of housekeepers on a given day is insufficient, the managers fill in from the "sister" areas. Finally, this arrangement has created an environment wherein employees do not harbor jealousies or inflated egos regarding their positions versus related slots in other hospitality-related areas. As a matter of policy, if an employee,

such as a foodservice worker, believes that the work is easier in housekeeping, he or she is encouraged to work in that area—after sufficient training, of course. Moreover, cross-training sessions (see the later discussion in this chapter) are commonplace, making such moves ordinary events.

The second unusual feature is the way employees are trained. Typically, training sessions involve members from all the hospitality services for two reasons. First, it results in effective cross-training. Employees gain proficiency in different skills, which directly aids in the deployment of resources. Second, the common training forum provides opportunities for employees to better understand the operation as a whole instead of viewing it myopically.

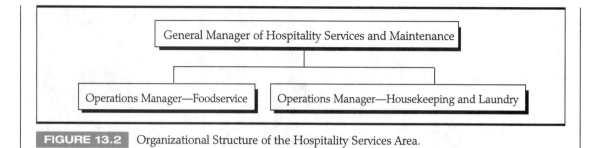

FIGURE 13.2 Organizational Structure of the Hospitality Services Area.

Has this innovative approach improved the level of service? According to third-party surveys, resident satisfaction is at an all-time high. The managers have also noted that employees support each other more than ever before. No longer do housekeepers openly criticize the food or treat the laundry-handling process as subservient work. Employees view themselves as part of a larger interdependent group instead of as isolated entities. The residents appreciate it. As one sister recently stated, "God is here with us. Even the employees understand that working together is the way He intended it to be."

What Makes a Good General Manager of Support Services?

It is fairly straightforward to outline what is needed when looking to hire a good food-service manager. This gets a little more difficult, however, when an organization seeks someone to run foodservice, housekeeping, and maintenance. Is it important to be a master in each area? Or is proficiency in any one area the most important attribute? If so, which one?

Most seasoned multiservice managers agree that while technical skills are a plus, human-resource-management skills are most critical. Next, as mentioned earlier, are time-management skills. Many managers, upon assuming responsibility for more than one service, find their days growing even longer, creating challenges for aspects of their lives that are not work related (such as marriages, families, social obligations, and personal fitness).

In addition to having good financial management skills, the other area that leads to success in general management is **project management.** Project management requires aptitude in planning, scheduling, managing resources, and tracking and controlling related costs during the finite term of a venture. As the number of services a manager oversees grows, project-management skills are even more important because they afford the manager the opportunity to build experience and demonstrate competence in areas beyond what is already listed on the résumé.[2]

Finally, general managers must, above all else, be good communicators. Since the **span of control**—the number of people the manager oversees—is so great, high-quality communication sends a clearer common message. In addition, general managers must be able to communicate with subordinates and superiors with the same level of confidence and efficacy. As though this list of requirements wasn't already long, a good general manager must also communicate well with large groups of people and must be able to express ideas in writing in such a way that others can easily understand the central message.

With these attributes in place, the only critical success factor remaining is a good fit with the organization. Even the most talented general managers have idiosyncrasies. In addition, every manager has developed unique ways of conducting everyday business. It is imperative that these attributes blend with the organization's culture and general practices.

CROSS-UTILIZATION OF LABOR

The efficiencies and labor savings cited earlier should be a given when hospitality services are combined. Fully leveraging the synergies, and cross-utilizing labor completely, however, require particular human-resource-management skills and practices. Building on the discussion from Chapter 8, these include a holistic approach to orientation and cross-training that is both elemental and developmental.

Holistic Orientation

Orientation in most organizations, at least for line employees in support-services functions, involves completing paperwork, meeting the superior to whom the person will report (if that person was not involved in the interview process), and perhaps having a tour of the departments. Too often this process is rushed, with the tour amounting to little more than directions to the restroom. Given the complexities of an environment that includes multiple services, this can be incredibly overwhelming and counterproductive.

In stark contrast to this, consider the **holistic orientation,** in which a dedicated orientation program socializes new employees to the workplace. This socializing serves partly as training, partly as orienting, and partly as directing. For multiservice operators, this process is divided into two parts: (1) host-organization orientation and (2) departmental and job orientation. Granted, this two-pronged approach is a good practice in any setting. But in a multiservice environment, the demands on and expectations for employees are particularly diverse, making an enveloping orientation that much more important.

Host-Organization Orientation   Regardless of the setting, whether it is a university, a prison, a corporate headquarters, or any other core business that the hospitality services support, every employee needs an **orientation to the host organization**—its products, mission, history, and market position. If a new employee is expected to contribute, even by cleaning bathrooms and washing dishes, it is imperative that he or she understand what the organization does. How else can management expect the employee to feel a part of the organization?

This general orientation should include aspects that are central to the host organization, such as its obligation to produce a profit while being socially responsible. We must, however, regard this point with caution. As indicated in Chapter 8 regarding the importance of honest, realistic job previews, information presented during orientation must also be realistic. Nothing impacts the first impression of a new employee as negatively as being sold a bill of goods during orientation, only to hear from other employees that the information is simply a deluded sales pitch.

Perhaps more relevant to the new employee, host-organization orientation should also include the nuts and bolts of pay, benefits, and specific terms and conditions of employment. This information is usually best delivered in pieces, intermixed between sessions about the host organization. In this way, employees can digest the information most easily and begin to feel more comfortable in their new setting. This balance of information also usually leads to more positive outcomes and perceptions.

Departmental and Job Orientation   Given the complexities of a multiservice structure, particularly one in which employees are expected to transcend traditional areas of responsibility, a thoughtful orientation to the various facets of the department (including whatever functions are part of the hospitality suite of services) is paramount.

Many human-resource management experts suggest that the first meeting with supervisors and coworkers is one of the most important ones. A rushed, seemingly impromptu meeting sends the message to new employees that they are not important. In stark contrast, a meeting that is structured, and in which management is well represented, translates into a cooperative work environment that the new employee can be expected to foster.

Departmental and job orientation must also introduce employees to their essential tasks and duties, both universal across support services and unique to the specific on-site operation. This is not to say that new employees should be thrust into the job with a slap on the back and an assurance that the manager on duty will check on their progress later. Rather, this means outlining exactly what tasks the job entails. Even for seasoned employees who have worked in one or more support services in the past, the concept of combined services is not easily swallowed at first blush.

Following this part of the orientation program, employees should then be introduced to relevant supervisors and coworkers. This allows management to properly shape the initial introductions. The associated benefit is that it sends an implicit message to existing employees that they, too, are important and that orientation is not just about hiring new employees. In truth, it is about building and constantly renewing the operation. This relates back to the initial discussion about the importance of internal customers in Chapter 7; while customers are vital to on-site operations, employees too are critical to the entire service enterprise. This aspect of the orientation also allows an opportunity for management to pair each new employee with a mentor, preferably a good, seasoned employee. Through ongoing **mentoring,** the veteran employee can provide one-on-one support regarding both the process-oriented and social aspects of the job.

The final part of the multiservice department orientation is a formal introduction to the specific social characteristics of the department's culture. This is probably the most often-ignored aspects of orientation, yet it is one of the most vital in indoctrinating new hires. Norms, values, symbols, and rituals take considerable time to establish within an on-site environment. Failing to introduce new employees to these needlessly compromises the work that management devotes to creating them in the first place.

Obviously, this is a lot of information to cover, which equates to both time and money for the employer. Again, however, it is important to think of an employee's orientation as a training session. Thus, the resources equate to money well spent.

How much time should be devoted to an orientation program, at least at the departmental level? There is no one right answer. The key to success is not to bombard new hires with information in a series of day-long orientation sessions. They will likely retain little of what is discussed and will quickly feel overwhelmed. Instead, well-thought-out, effective orientation programs, divided into shorter modules, say two hours long, are typically spread over several days. Again, it is important to note that these programmatic recommendations are advisable for any employee. Yet in a multiservice setting, the amount of information is so great, particularly given the background and experience of most support-service new hires, that thoughtful, planned delivery of orientation information is critical.

On the first day—the only time the employee spends all day in orientation activities—the new hire experiences the host-organization orientation and the first module of the departmental orientation, which should include the information that is most relevant and needed immediately. While full, the day is usually broken up nicely by breaks for the employees to complete paperwork and enjoy meal periods.

After that, an orientation program in a multiservice department should span several days, again with each day involving only two hours or so for orientation and the rest for working in the related position. This allows management to introduce different aspects of the job intelligently, such as foodservice-related responsibilities on one day and housekeeping issues on another. It also allows time for management to provide ample information about the human side of the work—all the more important in hospitality-related services. Optimally, the employee will have received not only a functional orientation, but also training in the multiple services within the first week.

The final critical part of orientation, which is particularly important in multiservice settings, is follow-up and evaluation of the orientation process. As noted earlier, the orientation, as outlined above, is not cheap; the need to monitor its effects is therefore that much more important. Throughout the process, managers should routinely check on new hires to ensure that information is flowing as planned. At the conclusion of the program, the new employees should be queried about what was useful, what seemed unimportant, and what might have been done better. Again, this practice communicates to employees that their input is important. It also provides invaluable information for general managers to continually refine the orientation process.

Cross-Training

The benefits and importance of **cross-functional training** (commonly referred to simply as **cross-training**) have been touched on in previous chapters, primarily in the context of maximizing productivity by expanding job responsibilities (see Chapters 8 and 9). In a multiservice environment, however, cross-training is the one tool that is essential to success in merging multiple services. In fact, if cross-training is not embraced fully, there is little reason to strive for efficiencies beyond those gained from flattening the management structure.

So what is cross-training? By definition, it is the *training of employees leading to the performance of duties and operations beyond their initially assigned jobs.* Relating back to Chapter 8, this extends the idea of job rotation more fully, encompassing more areas of a multiservice department.

Five basic steps are common to almost all cross-training efforts. The first of these is preparation for learning the new job. This requires an employee who wants to learn, since management may otherwise have considerable difficulty motivating an employee to take on new, often complex, tasks. Moreover, employees who lack sufficient desire to learn new tasks will likely require more subsequent training sessions than those employees who are ready to accept the challenge.

It should be noted that simply showing interest in the employee and explaining the importance of the new job can influence the employee's desire to learn. To that end, it is useful for a manager to review the targeted employee's past experience in order to properly frame the introduction to the new task. Again, this familiarity with the employee's experience communicates the idea that management is interested in the individual's development.

The second step involves reducing the new job to specific components and critical control points. In each area of multiservice departments, various tasks have a series of natural linkages, with junctures that represent these control points; it is all a matter of analyzing the job appropriately. Another important part of this step is identifying and communicating "tricks of the trade" that lend themselves to accomplishing various aspects of the task more easily. These not only help the employee to complete the new job more artfully, they also help the employee remember various steps in the process.

Artful presentation of the job, broken down into its components, is the next step. Simply telling an employee how to perform a task is not sufficient. Consider trying to teach an employee how to fold a cloth napkin in the shape of a tulip. Certainly a manager could describe the process, even breaking it down into its minutest steps. Yet without a demonstration, the employee will have considerable difficulty executing the task.

This leads to the fourth step, which is actual practice by the employee. Continuing our example of napkin folding, this step equates here to the employee's actually doing the task while following the trainer's instructions. Through basic trial and error, the employee can then master the simpler tasks and gain confidence regarding the more complex ones.

The final step is follow-up. As in most employee-related initiatives, cross-training requires monitoring, encouragement, and periodic retooling. An added benefit of follow-up is that many employees find ways to complete recently assigned tasks more efficiently than had been possible working in the manner in which they were trained. By monitoring subsequent activities, management can learn and share these efficiencies with other employees, leading to a natural continuous-improvement process.

Clearly, cross-training employees from several once distinct departments is a monumental task, yet one that is a natural part of combining multiple services. The good news is that employees generally gain momentum in learning diverse job functions. In addition, the desire to learn and gain new abilities increases with rising employee involvement. The outcome, of course, is a more talented, more productive staff that possesses a wide range of skills pertaining to multiple tasks.

One multiservice management team that has effectively harnessed these efficiencies has also created an interesting way of managing its diverse workforce, despite the hefty size of the host organization and the union environment in which it operates. Their approach is profiled as the chapter's final best practice.

BEST PRACTICES IN ACTION

GREATER BALTIMORE MEDICAL CENTER EMBRACES TECHNOLOGY AS A MEANS OF INTEGRATING MULTIPLE SERVICES (TOWSON, MARYLAND)

It's 8:46 A.M., and the emergency room has just called the nursing station in the east wing; a patient is on his way. The nurse checks and sees that a patient was just transferred from room 3823, which is at present the only vacant bed. Without worrying, the nurse calls the central "hospitality services station" and asks that the room be cleaned right away.

The computer operator who is staffing the station reports that a patient-service representative, George, is doing routine stocking of food in a nearby galley kitchen. The operator keys the room number that needs cleaning and George's code into the system. He also tells the system that the task needs immediate attention. The instant the information is entered, the signal is automatically transmitted to George's alphanumeric pager.

Seeing that room 3823 needs his attention, George gathers the necessary cleaning supplies from his workstation and enters the room, where he first uses the in-room phone's keypad to indicate to the computer that he is beginning the task. Once the room is clean, George uses the floor phone to tell the computer system that the task is complete and that he is returning to his earlier task.

Sound more like science fiction than reality? At the Greater Baltimore Medical Center, located in Towson, Maryland, events such as this occur routinely dozens of times each day.

Several years ago, the organization began looking for greater efficiencies in all of its functional areas. This is when the foodservice, housekeeping, and patient transportation departments were melded into a cohesive entity. Today, the multiservice area is very different from the traditional approach wherein these services operate in organizational silos. Staff members, instead of being allocated to an area such as housekeeping or foodservice, are instead scheduled by floor or wing.

This is made possible by an integrated support-services technology that tracks, organizes, and dispatches assignments to patient-service attendants (PSAs). The advantages are impressive. First, it allows the organization to practice patient-focused care (as discussed in Chapters 2 and 5) at an entirely new level. Each PSA has the opportunity to get to know patients. It is not unlikely, for example, that a PSA might help a patient make menu selections for the next meal and then transport him or her for a physical therapy session; the PSA might then clean the patient's room, perhaps change the light bulb that burned out earlier in the day, and then return to bring the patient back to the room. Later, the PSA might get a lemon wedge for the patient's tea and would likely clear the tray after the meal. This interchange creates a wonderful opportunity for the PSA to identify with the patient and to establish a relationship that leads to a higher level of care.

Another advantage is the analyses that are made possible by using this approach in concert with the technology. For example, productivity can be measured in a variety of ways using any number of partial-factor statistics (as discussed in Chapter 9). Comparisons can also be made at the most basic levels. For example, management can track how long it takes each employee to clean patients' rooms. The system also allows for the identification of areas of the institution that may become bottlenecks, such as rooms that are particularly difficult to clean or patient services that impair the timely transportation of patients.

Thus, the system allows for impartial monitoring of employee activity. Since employees log in before and log out after each task, it is simple to verify what employees are doing and when. Given the highly unionized environment, this is helpful in coping with the human-resources management challenge.

In sum, this novel approach leads to a higher level of care, is more efficient, allows for accurate tracking, and provides opportunities to manage better. Clearly, the approach as delivered by the team at the Greater Baltimore Medical Center is a best practice in the area of managing multiple services.

In conclusion, it should be noted that growth in managing multiple services stems primarily from contractors. As alluded to in Chapter 1, contractors market their expertise, inviting host organizations to outsource. It is a natural progression for contractors to look for opportunities beyond a single service offering. Nonetheless, the efficiencies, economies, and improvement in overall service are possible whether a suite of services is outsourced or performed in-house. The deciding factor, as in most cases, is the ingenuity, creativity, and ability of the managers.

CHAPTER SUMMARY

The advantages of merging two or more hospitality-related services include the savings in labor dollars resulting from a more parsimonious management structure and enhanced delivery of services gained from the reduction in organizational barriers. Another advantage is in the savings and efficiencies gained from sharing human resources among the operating areas.

Combining services, however, is logistically difficult and requires substantial planning and continuous review. It creates uncertainty for employees. It also changes the way general managers interact with employees; this is necessitated by the scope of the multiservice management role. Finally, the biggest challenge for those managing multiservice departments is the burden of limited time and abundant responsibility.

The manner in which services are combined reflects the needs that are typical of a particular host organization. The most common approach is to combine foodservice and housekeeping. Other logical additions to the hospitality-services area include laundry and maintenance. There are no limitations, but care must be taken to combine those elements that produce maximum opportunities for inter- and intradepartmental synergy.

The management skills needed to run a multiservice department begin with strong human-resource-management skills, followed by strong time-management skills. The ability to serve as project manager is also a plus. The capstone skill is the ability to communicate articulately and effectively, regardless of the size or hierarchical level of the audience.

In order to maximize the benefits of combined hospitality services as they apply to cross-utilization of labor, two elements are critical. The first is a holistic approach to orientation, which includes both a thorough host-organization orientation and a departmental and job orientation. The second is cross-training, which should be ongoing and pervasive within the multiservices department.

KEY TERMS

time management
project management
span of control
holistic orientation

host-organization
   orientation
departmental and job
   orientation

mentoring
cross-functional training
   (cross-training)

REVIEW AND DISCUSSION QUESTIONS

1. How did the practice of combining multiple services begin? In your opinion, will this trend continue?

2. Several advantages were noted pertaining to multiple services as an add-on to on-site foodservice. Can you think of other advantages? How about disadvantages other than those mentioned?

3. Name three ways in which time-management skills lead to better overall managerial prowess.

4. Think of an on-site operation that currently runs within the traditional organizational framework. Following a description of the host organization, outline the ideal combination of services and describe the corresponding management structure that you would recommend.

5. Do most organizations use a holistic-orientation approach? Why is this? Even if an on-site foodservice operation is not part of a multiple-services arrangement, might this approach still add value? Explain your response fully.

6. How can mentoring be applied in on-site foodservice beyond the approach described in this chapter, regardless of the orientation methods used?

7. Why is cross-training so vital to multiservices management? Does it also have a place in on-site foodservice when hospitality areas operate independently?

8. If you were the chief operating officer in charge of a host organization, say in a B&I setting, would you advocate combining the on-site foodservice department with other hospitality-related services? Why or why not?

END NOTES

[1] A good article that discusses time-management skills is Berger, F., & Merritt, E. (1998). No time left for you. *Cornell Hotel and Restaurant Administration Quarterly, 39*(5), 32–41.

[2] Two good sources for project management are Cotts, D. G., & Lee, M. (1999). *The facility management handbook.* New York: AMACOM; and Baker, S., & Baker, K. (1998). *Complete idiot's guide to project management.* New York: Alpha Books.

CHAPTER 14

EMERGING TECHNOLOGIES

Technology contributes to a better quality of life by making things easier. It does this by freeing people from mundane tasks and simplifying or optimizing others. And, of course, technology is alluring because of the fantastic possibilities it promises. This is certainly true in the foodservice industry, where problems are seemingly endless in number but solutions are rare—and sometimes unobtainable with current tools. Thus, this final chapter looks at technological prospects for on-site foodservice operators that may be a reality in the near—or perhaps more distant—future.

First, we consider emerging trends in hardware and software. Much has been said about this area in previous chapters, but this discussion brings to bear what will soon be part of tomorrow's best practices. The next topic is equipment. Ranging from robotics to equipment fabricated with a polymer that facilitates the creation of a bacteria-free environment, the products that incorporate new technology are nothing short of amazing. Finally, the chapter looks beyond the immediate horizon and extends new ideas from the drawing board to the operation of the future.

There are no best practices in this chapter, since much of the technology that is profiled has not hit the mainstream. It is hoped, however, that future editions of this book will feature many of the technologies viewed today as "emerging" with the same casualness that we view the basic point-of-sale system today.

NEW DEVELOPMENTS IN HARDWARE AND SOFTWARE

Thus far, we have profiled numerous applications that facilitate substantial operational improvements, whether in terms of food-production efficiencies, improved labor deployment, or enhanced guest satisfaction. It appears, however, that market-driven product development will continue to enhance the manner in which on-site foodservice operators can embrace such technology well into the foreseeable future. The greatest advances in hardware and software combinations will likely stem from two areas: wireless technology and Internet-aided systems.

Wireless Technology

The advantages of using wireless technology to enhance foodservice operations are already obvious. For example, employing handheld computers to take patient menu orders (as discussed in Chapter 5) or to enter inventory information (as discussed in

Chapter 7) is a useful application of today's technology and reduces labor costs while increasing customer satisfaction, employee productivity, or both. These tools, however, merely open the door to what is possible.

First, foodservice managers will soon develop a healthy reliance on real-time information about what each employee is doing and enjoy the associated benefit of being in contact with employees, regardless of where they work in an operation. The possibilities are already apparent, as evidenced by Chapter 13's second best practice, which showed how the foodservice management team at Greater Baltimore Medical Center tracks each employee's actions. This will be accomplished through the same technology that is available with some cell phones today, which include elements of **global positioning systems.**

In a large college campus setting, this could be invaluable. Need to notify the catering supervisor that 50 extra seats are required at the luncheon? No problem. Through the automated tracking system, you know she is en route with the ice sculpture. Simply call her on her cell phone. In turn, she sees that two employees are still setting up the room; she notifies them, and the job is already under way.

Such a system also offers managers rich methods of enhancing productivity. Through analytic tools built into the software, operators can evaluate what tasks require inordinate amounts of time and which employees are better at reducing such bottlenecks. Employees, too, will benefit. Never again will an employee claim ignorance as to how to deal with an unusual issue. Any employee, at any time, can use the communication tools to access supervisory support.

A related advantage of such technology is that it allows managers to better manage key interaction periods such as the lunch rush time. Commonly referred to as **moments of truth,** these times dictate how customers evaluate the overall operation. If managers can quickly evaluate whether staffing is sufficient given the volume of business (as evidenced by the point-of-sale tracking system), they can respond quickly in cases where a deficiency is noted.

The second likely application of wireless technology is in the enhancement of communication and task deployment. Envision a foodservice operation where an employee, upon clocking in, picks up messages regarding the day's duties. Cooks can quickly find out what is on the production schedule. Supervisors can learn if there were any callouts or special events for the day. While arguably lacking the human element, sharing information via wireless technology along with time-management and communication software enhances efficiency considerably.

Does this "big brother"-style tracking and depersonalized communication reduce foodservice management to an automated job void of humanity? Those with a disdain for technology might say yes, but those who are looking for ways to do a job better, and ultimately to enhance the guest's experience while increasing profits, will likely see the advantages of some of these tools of the future.

Internet-Aided Systems

Nearly every business guru and most visionaries cite the Internet as a critical element of every individual's future. Certainly, the Internet has already changed the way many things are accomplished from both a personal and a business perspective. Its pervasiveness is amazing, particularly as the word "Internet" garnered its now-colloquial definition very recently. Take, for example, estimates that members of the emerging **generation Y** (people born between 1981 and 1990) will spend close to a third of their lives—over 23 years on average—using the Internet in some capacity.

For on-site foodservice managers, network-aided systems that will play the largest role in the near future will relate to either marketing or purchasing and will transcend the operation in a variety of ways. Today, larger operators use the World Wide Web to post weekly menus and daily specials. Some, however, are already realizing the advantage of integrating this information with marketing programs.

For example, it is possible to post copies on a dedicated web site, updating specials accordingly. The advantage here is that the limitations of a paper menu are removed; only the skills and creativity of the operator limit the use of colors, graphics—including actual photos of some dishes—and animation. Thus, marketing can meld with merchandising, leading to greater penetration of the target market.

Through integrated technology, marketing via the Internet or intranets can also lead to services that today require employee involvement. It is reasonable to expect that new systems will allow an on-site operator to post a variety of catering menus, giving the customer the opportunity to customize at will. Upon entering specific information, the customer can then book an event. The system will automatically manage the number and type of events, ensuring that capacity in terms of rooms, labor, and production is not overextended.

Supply-chain integration and management will also change the way on-site operators perform the multitude of purchasing functions, thanks largely to systems that rely on Web-based technologies. Some operators already are realizing some of the efficiencies made possible by such applications. Florida Hospital's perpetual inventory system—a best practice featured in Chapter 7—already integrates the Web by allowing different operations to place orders from the centralized production facility. In the future, this same approach will integrate all facets of the perpetual inventory system, including the automation of production schedules that match sales levels.

Today, many large vendors have systems in place that allow on-site operators to place orders, giving information to both buyer and supplier regarding needs and product availability. The problem is that these systems are not integrated with other food-management systems used within the operation. In the very near future, operation-management systems will integrate **e-purchasing**—purchasing done on-line and in real time—with a diverse set of vendors. In turn, purchasing and receiving will be linked to inventory systems, facilitating much more accurate tracking. Finally, integration with point-of-sale systems will allow operators to maintain optimal perpetual inventory systems with a minimum of labor—less even than what is needed today for basic inventory-management approaches.

EQUIPMENT ON THE CUTTING EDGE

In 1988, K. G. Englehardt, the head of the Health and Human Services Robotics Laboratory at Carnegie Mellon University, introduced a robotic "arm" designed to package silverware in plastic bags in airline and healthcare operations. The invention was both lauded and despised. Foodservice operators, particularly those in the on-site segment, recognized the potential for enhancing productivity and relieving workers of repetitive jobs. Some unions and labor activists, however, viewed the robot as an unwelcome invader, greeting the banners touting the new technology with repugnance.

Englehardt's arm portended an era of automation that would alleviate many of the problems of managing unskilled workers, thereby allowing foodservice operators to focus on training and developing staff members who could directly enhance customers' dining experiences. Unfortunately, the field of robotics is only slowly entering the foodservice realm. Nonetheless, the innovation provides a natural introduction to a discussion of cutting-edge equipment, the most notable of which pertains to back-of-the-house processes and tools geared toward enhanced sanitation.

Production-Related Equipment

Commercial food-production equipment that is emerging today and tomorrow shares two key characteristics. First, it is functionally flexible and spatially versatile. For example, a leading-edge oven technology, dubbed **rapid cook,** combines microwaves with hot-air impingement. The air is directed toward the cooking surface rather than just circulating throughout the oven cavity, as is common in convection ovens. This technology allows a single piece of equipment to be used for multiple cooking approaches. Moreover, the rapid cook ovens that are being proposed occupy minimal space.

Another example of flexibility and versatility involves induction technology. As alluded to in Chapter 4 in discussing versatile marketplace-style eatery configurations, induction-heating surfaces do not require dedicated hoods. Perhaps of greater importance, they offer all of the benefits of cooking with gas and none of the drawbacks.

Induction equipment uses electromagnetic energy to heat cookware fabricated from nondielectric material, which may also have a nonstick cooking surface. When engaged, the internal coils produce a high-frequency alternating magnetic field that ultimately flows through the cookware. Since the molecular vibration in the cookware causes the heat, the cooktop's glass-ceramic surface remains relatively cool because it contains no magnetic material; only the heat from the pan warms the glass. Thus, induction cooktops are generally safer than traditional approaches, and provide considerable ease in cleaning since there is no chance of spills seeping into the burner or other crevices common in traditional cooktops. This technology has been in use for some time, yet the combination of induction heat with a variety of designs (e.g., display cooking, small side stations) will likely emerge as a technology that only increases in value.

The second characteristic common to all of tomorrow's food-production equipment is that it is more cost effective and energy efficient. The aforementioned induction equipment, for example, is less expensive to operate than gas or electric burners owing to its design. Furthermore, it uses full power only when there is a magnetic load to be heated.

New technology in "cook and hold" ovens, which will incorporate integrally wired probes that automatically regulate the cooking time of a product, is also financially beneficial. First, the ability of these ovens to cook at lower temperatures equates to a higher yield (as compared, say, with roasting meats at higher temperature). Even with longer cooking times resulting from the lower temperatures, the units are more efficient in terms of electricity usage. Finally, the probes allow for a more consistent end product because, once the item is done, the oven automatically switches to a "hold" state, thereby preventing overcooking yet ensuring safe holding of the cooked item.

The final example of how cutting-edge equipment can lead to better outcomes is evident in the next generation of robotics. While Englehardt's arm was somewhat limited in application, more recent robotic devices are capable of delivering food orders, scrubbing the kitchen floor, or "manning" the grill. An example of an automated grill cook, a recent introduction to the foodservice equipment realm, is a robot that can cook on a griddle, fryer, or steamer (see Figure 14.1). It delivers consistently prepared food items and ensures that items are cooked to a predetermined internal temperature, ensuring the utmost in food safety. While the utility of this latest product is yet undetermined, it is clear that the long-term savings in labor costs can be substantial. Moreover, the robot doesn't cost more if scheduled to work overtime—and it never complains.

Better Sanitation Through Technology

Food safety concerns everyone. And while industry standards and training dedicated to food-safety issues within the foodservice industry are at an all-time high, severe problems associated with employees' improper handling of food remain. The record

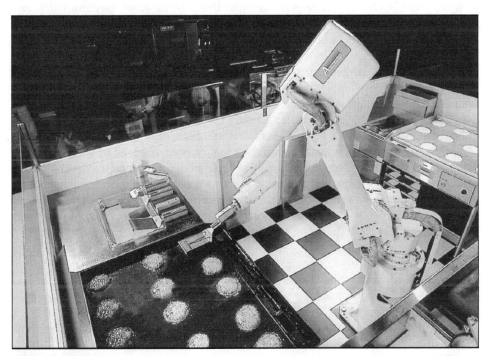

FIGURE 14.1 Accutemp's Flipper the Robocook, Shown Cooking Burgers and Pancakes. Copyright by Accutemp, Inc.

number of foodborne-illness-related incidents, including approximately 76 million illnesses, 325,000 hospitalizations, and 5,000 deaths in the United States in 2000, is evidence of this.[1]

Manufacturers have always attempted to integrate concerns for food safety in designing equipment. They have also underscored the importance of this in marketing their wares. For example, one of the key attributes of the robotic grill cook described earlier is the automated mechanism that ensures that items are cooked to a proper internal temperature.

This focus on food safety will continue and will be enhanced through new technology. For example, some equipment now entering the mainstream features a unique technological advancement that is based on the active ingredients found in today's popular antibacterial soaps, shampoos, and deodorants. Called Microban®, the technology integrates a combination of ingredients directly into the polymer's molecular structure. Thus, the equipment itself—not just a coating that may wear out over time—possesses the odorless, tasteless, and colorless ingredient that neutralizes the ability of microbes to function, grow, and reproduce. And because these ingredients are uniformly dispersed, the built-in antimicrobial benefits are present throughout the piece of equipment, even in hard-to-clean areas.

One piece of equipment that is often ignored but harbors potentially substantial foodborne-illness dangers is the icemaker. Many a foodservice operator has been embarrassed by a health inspector who examines an operation's ice machine and finds mold inside. In addition, the problems associated with moving ice from the icemaker to other areas, such as soda dispensers, open the door for cross-contamination and employee injury when lifting the ice.

The problem of keeping the internal workings of the machine cleaner will be solved in part with new technology and design. For example, the latest in icemakers create cubes by injecting streams of water through high-pressure nozzles into small, chilled cube-size cups; the cups are grouped in large trays, which are tiered so that water cascades through and is then recirculated until frozen. The seemingly wasteful process of recirculating excess water actually results in more appealing crystal-clear cubes and also maintains a higher level of cleanliness within the machine.

To reduce problems associated with moving ice from one area of an operation to another, an innovative piece of cutting-edge equipment is designed to prevent human contact with the ice or the surface on which the ice rests while reducing its exposure to airborne contamination. Such units can be located in remote areas of the kitchen and automatically deliver ice to stations up to 250 feet away by means of intense blasts of air and flexible tubes. The latest system of this type is capable of moving more than 600 pounds of ice per hour.

Regardless of the type of equipment, an employee's hands are the biggest source of foodborne contamination. Thus, proper hand washing has always concerned on-site operators. Again, technology promises to make this challenge easier. Some new systems automatically start the flow of water, dispense the proper amount of soap, and activate a dryer; some systems even generate reports on who washed, for how long, and how often.

The problem is that there is no way to monitor whether employees cleaned under their fingernails or between their fingers. A pioneering solution, stemming from

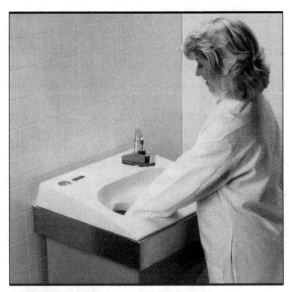

FIGURE 14.2 An Automated Hand-Washing System. © by Meritech Inc.

a cross section of industries, is an automated hand-washing system designed to remove harmful pathogens and contaminants from bare (as well as gloved) hands. With a series of rotating cylinders and sprayers, the system is extremely effective and requires only passive involvement of employees. All they have to do is put their hands fully into the machine and the technology does the rest. (For an example, see Figure 14.2.)

Finally, the last area of equipment that is designed with food safety in mind maps and maintains consistent temperatures. For example, walk-in refrigerators are being designed with internal tracking mechanisms to ensure that temperatures do not fluctuate outside a prespecified range. Similarly, fryers and griddles will soon track and adjust internal temperatures, leading to better, more consistent, and safer food production.

While these advances are impressive, the add-on feature of radio-frequency technology makes them more remarkable. Already, some manufacturers are talking about refrigerators and freezers that will transmit a signal to a cell phone or pager if temperatures rise above safe limits. The transmitted message will include the time and duration of the occurrence while also generating a log of the incident and the transmission. It may be hard to imagine, but managers may soon have to deal with calls not only from employees but also from the equipment they use!

BEYOND THE HORIZON

Each of these innovations is impressive. But as with all technology, these examples only hint at what might be possible. And given the amazing rate of advancement in the past few decades, the promise of the future is even more fantastic.

What will the distant future hold in terms of technology in on-site foodservice? It is expected that the trends discussed previously will continue. As today's technology becomes more a part of everyday living, however, features such as equipment that communicates with operators will be more transparent. Digital printouts will likely be replaced by audio—and possibly video—transmissions. It is not unthinkable, for example, to envision a dish machine that, rather than using lights to indicate that the sanitizer is low, will instead trigger a communiqué to the dish-machine operator, followed by a step-by-step video directive on how to refill it.

While such "Jetson"-like advances are conceivable, others are less easy to accept. For example, is it possible that a retail foodservice outlet will not include a cashier? After all, someone has to be there to accept payment for goods, right? In reality, this is a very likely scenario. Through intelligent cash registers and the likely prevalence of commonplace cashless-payment systems, the traditional staffing of a cashier in grab-and-go operations will likely go the way of the cigar box when electronic cash registers became affordable.

The seamless integration of software will also become much more prevalent. Even by today's standards, it's ridiculous that most on-site operations use one system to place orders, another to track inventory, and still another to manage labor records and payroll. Integrated systems will lead the way to even greater efficiencies and will enhance management's ability to focus on the nuances of effective leadership and motivation instead of paper shuffling and number crunching.

Finally, the future will most certainly bring a greater consolidation among food-service providers, distributors, and food and equipment manufacturers. This is the nature of a global economy working within a capitalistic environment. With this consolidation will come greater emphasis on unit-level efficiencies and shared economies. As intimated throughout this book, this will lead to an even more urgent need for managers to be adept in every facet of foodservice-operations management.

Concluding Thought

Clearly, the possibilities are impressive. Yet they present us with a paradox. The practice of foodservice management—whether art, science, or both—has always been predicated on people and the services they deliver. Recalling the discussion in Chapter 1, it has been this way for several millennia. The majority of emerging technologies, however, carry the implicit—and in many cases explicit—message that they can reduce labor and, in most instances, reduce the face-to-face contact that makes the hospitality industry the wonder that it is. The example of the grab-and-go eatery without a human cashier underscores this trend.

Will we eventually rework, retool, reengineer, and advance technologically to a state where human capital is all but wholly devalued? The answer is "No." Technology will advance, but with it people's expectations will continue to expand, just as we suggested in Chapter 4. True, there may be occasions when the human encounter is purposely diminished. This will likely be for the sake of convenience and speed of service. Yet the point of differentiation between one on-site operation and another will return again to where it has always been: service. And while machines streamline operations, service is—at least in foodservice—a product of the human element. Thus, for on-site foodservice managers, it will remain as Ellsworth Statler (1863–1928) expressed it so long ago: *Life is service; the one who progresses is the one who gives his fellow men a little more—a little better service.*

KEY TERMS

global positioning systems   generation Y   Englehardt's arm
moments of truth   e-purchasing   rapid cook

REVIEW AND DISCUSSION QUESTIONS

1. Several applications of wireless technology were discussed. Think of two others and describe how they might influence foodservice operations.
2. What are some of the implications—both positive and negative—of using global positioning systems to track employees?
3. Why are moments of truth important?
4. Describe how a menu should be featured when using a portal or web site. What are the key elements? Do these differ from those of traditional paper menus?
5. Why didn't Englehardt's arm produce dramatic labor reductions?
6. In your vision of the future, what will the back of the house look like? Describe the equipment, labor utilization, and product flow.
7. Given the challenges of hand washing, why doesn't every operation employ some type of automation, or at least tracking, of this vital sanitation practice?
8. Given the prediction that managers will be increasingly "wired" to their operations in the future, what does this mean in terms of separating work and home? Include both the potentially positive and negative aspects of this trend.
9. If you were asked to predict the next innovation in on-site foodservice technology, what would you say? Be as specific as possible.

END NOTES

[1] While reports surrounding specific issues of foodborne illnesses are numerous, these statistics summarize the impact of the problem. For more on the statistics, related concerns, and the effect of education, see Medeiros, L., Hillers, V., Kendall, P., & Mason, A. (2001). Evaluation of food safety education for consumers. *Journal of Nutrition Education, 33,* 27–34.

ABOUT THE AUTHOR

As a member of the Cornell University School of Hotel Administration's faculty, Dennis Reynolds teaches and conducts research on management topics relevant to the foodservice industry. His teaching—targeting undergraduate, graduate, and executive-education students—encompasses specialized niches of foodservice management including those serving the healthcare, corporate, B&I, and education sectors.

Professor Reynolds' research focuses on subjects related to managerial motivation, with particular attention to human-resource issues salient to the foodservice industry and issues related to productivity analysis and enhancement. One area of interest involves enhancing managerial efficacy through effective feedback mechanisms; another is the study of productivity measures using data envelopment analysis, thereby allowing for the formation of optimal efficiency frontiers.

In maintaining his connection with the industry, Dr. Reynolds works with a variety of leading hospitality companies in an effort to maximize their human capital. His consulting efforts share a commonality of maximizing operations' market penetration and profitability through a variety of programs, including efficiency analysis, revenue management, internal-control implementation, and quality assurance.

Prior to embarking on his career as an academic, Dr. Reynolds served in the corporate ranks with global hospitality-management companies. Most recently, he ran a division of a publicly traded contract foodservice company managing over $30 million in annual volume and more than 3,000 employees. Building on his college years in the restaurant-management trenches, he also spent several years with a leading consulting firm, servicing both the lodging and foodservice industries.

Professor Reynolds holds a doctoral degree from Cornell University in hospitality management and a master of professional studies degree with a concentration in internal controls, also from Cornell University, as well as a bachelor of science degree in hotel, restaurant, and institutional management from Golden Gate University. He has presented scholarly papers at various conferences and conventions, including those conducted by the Academy of Management and the International Council on Hotel, Restaurant, and Institutional Education. In addition to book chapters, his publications include articles in such journals as the *Cornell Hotel and Restaurant Administration Quarterly*, the *Journal of Service Research*, the *Advanced Management Journal*, the *Journal of Hospitality and Tourism Research*, the *Journal of Foodservice Business Research*, and the *FIU Review*. He serves on the editorial review board of the *Advanced Management Journal*, the *Journal of Human Resources in Hospitality and Tourism*, the *International Journal of Hospitality and Tourism Administration*, and *Praxis: The Journal of Applied Hospitality Management*.

Of greatest personal importance, Dennis and his wife, Julia, are proud parents of two wonderful girls. The Reynolds family currently resides in Ithaca, New York.

INDEX

Prisons, *see* On-site foodservice, correctional facilities
Process-based methods of quality, *see* Quality
Production systems:
  centralized food production, 42, 46, 48, 54, 231
  decentralized food production, 42, 54
Productivity, 7, 9, 45, 103, 143–155, 157, 158, 217, 223, 226, 230
Profit and loss (P&L) contracts, 19–20, 25
Profit generator, 100
Program implementation, *see* Training
Project coordination, 92, 101
Project management, 220, 227, 228
Purchasing, 103, 105–110, 120, 186
Pygmalion effect, 177–178, 181, 182. *See also* Golem effect

Quality, 31, 185, 187–200
  affinity process, 197, 199
  continuous quality improvement (CQI), 189-190, 199
  defined, 187
  customer-based measures, 188, 199
  detection-based methods, 188, 199
  financial-based methods, 188, 199
  hoshin planning, 196–197, 199, 200
  International Organization for Standardization (ISO), 188, 195, 199
  ISO9000, 195, 199
  Pareto diagram, 197–198, 199, 200
  Picos, 188, 191–194, 199
  process-based methods, 188, 199
  quality circles, 190, 199
  Total Quality Management (TQM), 178, 188, 189–191, 192, 196, 199
Quality circles, *see* Quality

Rapid cook, 232, 237
Raven, Bertram, 176
Ready prepared, 42, 54
Reagan, Ronald, 114
Rebates, 110, 120, 121
Receiving, 105, 110–111, 114, 120, 121
Recency bias, 134, 140. *See also* Interview mistakes
Réchaud, 53
Recognition programs, 150–151, 153, 154, 155. *See also* Productivity

Recreation facilities, *see* On-site foodservice, recreation facilities
Recruiting, 103, 123, 129–132, 139, 140, 157
Regional concepts, 69, 73
Referent power, *see* Power
Restaurant-style menu, 77
Restaurant-style service, 75, 77, 78–80, 84, 86, 87
Retention, 103, 123, 137–139, 158. *See also* Turnover
Rethermalization, 43, 49, 54
Retirement community, *see* Continuing care retirement community
Return on investment, 42–43, 116
Reward power, *see* Power
Reynolds, Dennis, 26, 121, 155, 183, 239
Richman, Phyllis, 16
Ries, A., 74
Ries, L., 74
Room service, *see* Hotel-style room service
Rothwell, W. J., 141
Rutes, W. A., 74
Ryan, M. J., 200

Sales mix, 150
Sales per labor hour, 144, 154
Scandling, William, 26
Scatter system, 62, 73, 74
Schools, *see* On-site foodservice, schools
Scratch cooking, 42, 54
Seashores, S. E., 183
Selection, 103, 123, 132–134, 139, 140, 157
Selection ratios, 131, 140
Self-operated foodservice, 20, 27
Self-service, *see* Service styles
Senge, P. M., 39
Senior living, 185, 201–214. *See also* Eldercare centers
  assisted living, 206, 207, 213
  continuing care retirement community, 72, 205, 208, 209, 210–212, 213
  home care, 206, 207–208, 213
  hospice care, 206, 207–208, 213
  independent living, 206–207, 213
  segments, 206–208
  skilled nursing, 206, 207, 210, 213
Service styles:
  portable meals, 52, 54
  self-service, 51, 54
  tableside preparation, 53, 54

Trait approaches, *see* Leadership
Transactional leadership, *see* Leadership
Transformational leadership, *see* Leadership
Transportation-related foodservice, 4
Travis, D., 74
Tray service, *see* Service styles
Trend-projection technique, 124, 140
Trigger, 105, 120
Tube feeding, 146
Tumble chiller, 43, 44
Turnover, 123, 129, 137, 138, 139, 140, 141.
    See also Retention
Turnover rate determination, 137–138

UCLA Medical Center, 136–137. *See also*
    Best practices
Umbrella branding, 70, 73, 74
Universities, *see* On-site foodservice,
    colleges and universities
University of Massachusetts Memorial
    Healthcare System, 44

Valence, 173, 174, 182. *See also*
    Motivation
van de Vliert, E., 183

Velthouse, B. A., 184
Vending, 50
Vertical dyad linkage, *see* Leader-member
    exchange (LMX)
Vroom, Victor, 173, 183

Wagner, C. G., 26
Wait service, *see* Service styles
Wayne, S. J., 183
Weighted averages, *see* Forecasting
Wendy's, 47
Wiersema, W., 121
Wireless technology, 229–230, 237
Woods, R. H., 141
Worker safety programs, 150, 152–153, 154.
    See also Productivity
Workflows, 129, 140, 193

Xerox Corporation, 188

Yield ratios, 131, 140
Yukl, G., 183

Zaleznik, A., 183
Zero-based budgeting, 31